TECHNICAL ENGINEERING AND DESIGN GUIDES
AS ADAPTED FROM THE
US ARMY CORPS OF ENGINEERS, NO. 3

Design, Construction, and Maintenance of
RELIEF WELLS

Published by
ASCE Press
American Society of Civil Engineers
345 East 47th Street
New York, New York 10017-2398

ABSTRACT

This engineering manual, *Design, Construction, and Maintenance of Relief Wells: Technical Engineering and Design Guides as Adapted from the US Army Corps of Engineers, No. 3,* provides guidance and information on the design, construction, and maintenance of pressure relief wells. The manual covers relief wells that are installed for the purpose of relieving subsurface hydrostatic pressures that may develop within the pervious foundations of dams, levees, and hydraulic structures. It begins with a discussion of certain basic considerations such as foundation investigations, anisotropic conditions, and seepage analysis. Next, the manual covers the analysis of single and multiple well systems followed by the design of these wells. Successive topics discussed are the installation, inspection and maintenance, malfunctions, and rehabilitation of wells. Finally, the engineering manual provides detailed information on the mathematical analysis of underseepage and substratum pressures.

Library of Congress Cataloging-in-Publication Data

Design, construction, and maintenance of relief wells.
    p.cm.—(Technical engineering and design guides as adapted from the U.S. Army Corps of Engineers; no. 3)
    Includes bibliographical references and index.
    ISBN 0-87262-955-4
    1.Relief wells—Design and construction. 2.Relief wells— Maintenance and repair. I.American Society of Civil Engineers. II.United States. Army. Corps of Engineers. III.Series.
TC973.D47 1993   93-43315
627—dc20   CIP

    The material presented in this publication has been prepared in accordance with generally recognized engineering principles and practices, and is for general information only. This information should not be used without first securing competent advice with respect to its suitability for any general or specific application.
    The contents of this publication are not intended to be and should not be construed to be a standard of the American Society of Civil Engineers (ASCE) and are not intended for use as a reference in purchase specifications, contracts, regulations, statutes, or any other legal document.
    No reference made in this publication to any specific method, product, process or service constitutes or implies an endorsement, recommendation, or warranty thereof by ASCE.
    ASCE makes no representation or warranty of any kind, whether express or implied, concerning the accuracy, completeness, suitability or utility of any information, apparatus, product, orprocess discussed in this publication, and assumes no liability therefor.
    Anyone utilizing this information assumes all liability arising from such use, including but not limited to infringement of any patent or patents.

Photocopies. Authorization to photocopy material for internal or personal use under circumstances not falling within the fair use provisions of the Copyright Act is granted by ASCE to libraries and other users registered with the Copyright Clearance Center (CCC) Transactional Reporting Service, provided that the base fee of $2.00 per article plus $.25 per page copied is paid directly to CCC, 27 Congress Street, Salem, MA 01970. The identification for ASCE Books is 0-87262-955-4/93 $2.00 + $.25. Requests for special permission or bulk copying should be addressed to Permissions & Copyright Dept., ASCE.

Copyright © 1993 by the American Society of Civil Engineers,
exclusive of U.S. Army Corps of Engineers material.
All Rights Reserved.
Library of Congress Catalog Card No: 93-43315
ISBN 0-87262-955-4
Manufactured in the United States of America.

TABLE OF CONTENTS

Chapter 1. Introduction

1-1 Purpose	1
1-2 Objective and Scope	1
1-3 Applicability	1
1-4 References	1
1-5 General Consideration	1

Chapter 2. Relief Well Applications

2-1 Description	2
2-2 Use of Wells	2
2-3 History of Use	2
2-4 Other Applications	4

Chapter 3. Basic Considerations

3-1 Foundation Investigations	6
3-2 Foundation Permeability	6
3-3 Anisotropic Conditions	6
3-4 Chemical Composition of Ground Waters	6
3-5 Seepage Analysis	11
3-6 Allowable Heads	11

Chapter 4. Analysis of Single Wells

4-1 Assumptions	13
4-2 Circular Source	13
4-3 Noncircular Source	13
4-4 Infinite Line Source	13
4-5 Finite Line Source	15
4-6 Infinite Line Source and Infinite Line Sink	15
4-7 Infinite Line Sink and Infinite Barrier	15
4-8 Complex Boundary Conditions	15
4-9 Partially Penetrating Wells	15
4-10 Effective Well Penetration	18

Chapter 5. Analysis of Multiple Well Systems

5-1 General Equations	21
5-2 Empirical Method	21
5-3 Circular Source	22
5-4 Wells Adjacent to Infinite Line Source with Impervious Top Stratum	22
5-5 Infinite Line of Wells	22
5-6 Top Stratum Conditions	22
5-7 Infinite Line of Wells, Impervious Top Stratum	22

5-8 Well Factors	23
5-9 Infinite Line of Wells, Impervious Top Stratum of Finite Length	24
5-10 Infinite Line of Wells, Impervious Top Stratum Extending to Blocked Exit	25
5-11 Infinite Line of Wells, Discharge Below Ground Surface	26
5-12 Infinite Line of Wells, No Top Stratum	26
5-13 Finite Well Lines, Infinite Line Source	28

Chapter 6. Well Design

6-1 Description of Well	35
6-2 Materials for Wells	35
6-3 Selection of Materials	35
6-4 Well Screen	36
6-5 Filter	36
6-6 Selection of Screen-Opening Size	37
6-7 Well Losses	38
6-8 Effective Well Radius	41

Chapter 7. Design of Well Systems

7-1 General Approach	42
7-2 Design Heads	42
7-3 Boundary Conditions	42
7-4 Design Procedures	42
7-5 Infinite Line of Wells, Impervious Top Stratum	42
7-6 Infinite Line of Wells, Wells in Ditch	43
7-7 Infinite Line of Wells with Impervious Top Stratum of Finite Length	43
7-8 Computer Programs	45
7-9 Head Distribution for Finite Line of Relief Wells	45
7-10 Well Systems at Outlet Works and Spillways	47
7-11 Well Costs	47
7-12 Seepage Calculations	47

Chapter 8. Relief Well Installation

8-1 General Requirements	49
8-2 Standard Rotary Method	49
8-3 Reverse-Rotary Method	49
8-4 Bailing and Casing	51
8-5 Bucket Augers	51
8-6 Disinfection	51
8-7 Installation of Well Screen and Riser Pipes	53
8-8 Filter Placement	53
8-9 Development	54
8-10 Chemical Development	54
8-11 Mechanical Development	54
8-12 Sand Infiltration	57
8-13 Testing of Relief Wells	57
8-14 Backfilling of Well	57

8-15 Sterilization 58
8-16 Records 59
8-17 Abandoned Wells 59

Chapter 9. Relief Well Outlets

9-1 General Requirements 60
9-2 Check Valves 60
9-3 Outlet Protection 60
9-4 Plastic Sleeves 60

Chapter 10. Inspection Maintenance and Evaluation

10-1 General Maintenance 64
10-2 Periodic Inspections 64
10-3 Pumping Tests 64
10-4 Records 65
10-5 Evaluation 65

Chapter 11. Malfunctioning of Wells and Reduction in Efficiency

11-1 General 66
11-2 Mechanical 66
11-3 Chemical 66
11-4 Biological Incrustation 67

Chapter 12. Well Rehabilitation

12-1 General 68
12-2 Mechanical Contamination 68
12-3 Chemical Treatment with Polyphosphates 68
12-4 Chemical Incrustations 68
12-5 Bacterial Incrustation 69
12-6 Recommended Treatment 69
12-7 Specialized Treatment 70

Appendices

Appendix A   References 71
Appendix B   Mathematical Analysis of Underseepage and Substratum Pressures 73
Appendix C   List of Symbols 83

DEPARTMENT OF THE ARMY
U.S. Army Corps of Engineers
WASHINGTON, D.C. 20314-1000

REPLY TO
ATTENTION OF:

Mr. James W. Poirot
President, American Society
  of Civil Engineers
345 East 47th Street
New York, New York 10017

Dear Mr. Poirot:

I am pleased to furnish the American Society of Civil Engineers (ASCE) a copy of the U. S. Army Corps of Engineers Engineer Manual, EM 1110-2-1914, Design, Construction and Maintenance of Relief Wells. The Corps uses this manual to provide information on pressure relief wells used to relieve subsurface hydrostatic pressures within the previous foundations of dams, levees and hydraulic structures.

I understand that ASCE plans to publish this manual for public distribution. I believe this will benefit the civil engineering community by improving transfer of technology between the Corps and other engineering professionals.

Sincerely,

Arthur E. Williams
Lieutenant General, U. S. Army
Commanding

CHAPTER 1

INTRODUCTION

1-1. Purpose

This manual provides guidance and information on the design, construction, and maintenance of pressure relief wells.

1-2. Objective and Scope

The objective of this manual is to provide guidance and information on the design, construction, and maintenance of pressure relief wells installed for the purpose of relieving subsurface hydrostatic pressures which may develop within the pervious foundations of dams, levees, and hydraulic structures.

1-3. Applicability

The provisions of this manual are applicable to all HQUSACE/OCE elements, major subordinate commands, districts, laboratories, and field operating activities (FOA) having responsibility for seepage analysis and control for dams, levees, and hydraulic structures.

1-4. References

Appendix A contains a list of required and related publications pertaining to this manual. Unless otherwise noted, all references are available on interlibrary loan from the Research Library, ATTN: CEWES-IM-MI-R, US Army Engineer Waterways Experiment Station, 3909 Halls Ferry Road, Vicksburg, MS 39180-6199.

1-5. General Considerations

All water retention structures are subject to seepage through their foundations and abutments. In many cases the seepage may result in excess hydrostatic pressures or uplift pressures beneath elements of the structure or landward strata. Relief wells are often installed to relieve these pressures which might otherwise endanger the safety of the structure. Relief wells, in essence, are nothing more than controlled artificial springs that reduce pressures to safe values and prevent the removal of soil via piping or internal erosion. The proper design, installation, and maintenance of relief wells are essential elements in assuring their effectiveness and the integrity of the protected structure.

CHAPTER 2

RELIEF-WELL APPLICATIONS

2-1. Description

Pressure relief wells as used in this manual refer to vertically installed wells consisting of a well screen surrounded by a filter material designed to prevent inwash of foundation materials into the well. A typical relief well is shown in Figure 2-1. The wells, including screen and riser pipe, have inside diameters generally between 6 and 18 inches (in.), sized to accommodate the maximum design flow without excessive head loss. Well screens generally consist of wire-wrapped steel or plastic pipe, slotted or perforated steel or plastic pipe. Slotted wood stave well screens, which are no longer manufactured, are found in many existing installations. Details of various well screens are given in Chapter 6.

2-2. Use of Wells

A. Relief wells are used extensively to relieve excess hydrostatic pressures in pervious foundation strata overlain by more impervious top strata, conditions which often exist landward of levees and downstream of dams and various hydraulic structures. Placing the well outlets in below-surface trenches or collector pipes serves to dry up seepage areas downstream of levees and dams. Relief wells are often used in combination with other underseepage control measures, such as upstream blankets, downstream seepage berms, and grouting. Horizontal stratification of pervious foundation deposits is not a major deterrent to the use of relief wells, as each of the more pervious foundation strata can be penetrated. The use of relief wells for levee systems is discussed in EM 1110-2-1913; their use for earth and rock-fill dams is discussed in EM 1110-2-2300.

B. Relief wells provide a flexible control measure as the systems can be easily expanded if the initial system is not adequate. Also, the discharge of existing wells can be increased by pumping if the need arises. A relief well system requires a minimum of additional real estate as compared with other seepage control measures such as berms. However, wells require periodic maintenance and frequently suffer loss in efficiency with time for a variety of reasons such as clogging of well screens by intrusions of muddy surface waters, bacterial growth, or carbonate incrustation. Relief wells may increase the amount of underseepage which must be handled at the ground surface, and means for collecting and disposing of their discharge must be provided (Turnbull and Mansur 1954). Adequate systems of piezometers and flow measuring devices must be installed in accordance with ER-1110-2-110 and EM 1110-2-1908 to provide continuing information on the performance of relief well systems.

2-3. History of Use

A. The first use of relief wells to prevent excessive uplift pressures at a dam was by the US Army Engineer District, Omaha, when 21 wells were installed from July 1942 to September 1943 as remedial seepage control at Fort Peck Dam, Montana (Middlebrooks 1948). The foundation consisted of an impervious stratum of clay overlaying pervious sand and gravel. Although steel sheet piles were driven to provide a complete cutoff, leakage occurred and high hydrostatic pressure developed at the downstream toe with an excess head of 45 feet (ft) above ground surface. The high pressure was first observed in piezometers installed in the pervious foundation. The first surface evidence of the high hydrostatic pressure came in the form of discharge from an old well casing that had been left in place. Since it was important that the installation be made as quickly as possible, 4- and 6-in. well casings, available at the site, were slotted with a cutting torch and installed on 250-ft centers in the pervious stratum with solid (riser) pipes extending to the surface. The excess head at the downstream toe was reduced from 45 to 5 ft, and the total flow from all wells averaged about 4500 gallons per minute (gpm). However, the steel screens corroded severely and in 1946 were replaced by 17 permanent wells consisting of 8-in.-ID slotted redwood pipe at a spacing of 125 ft.

B. The first use of relief wells in the original design of a dam was by the US Army Engineer District, Vicksburg, when wells were installed dur-

RELIEF-WELL APPLICATIONS

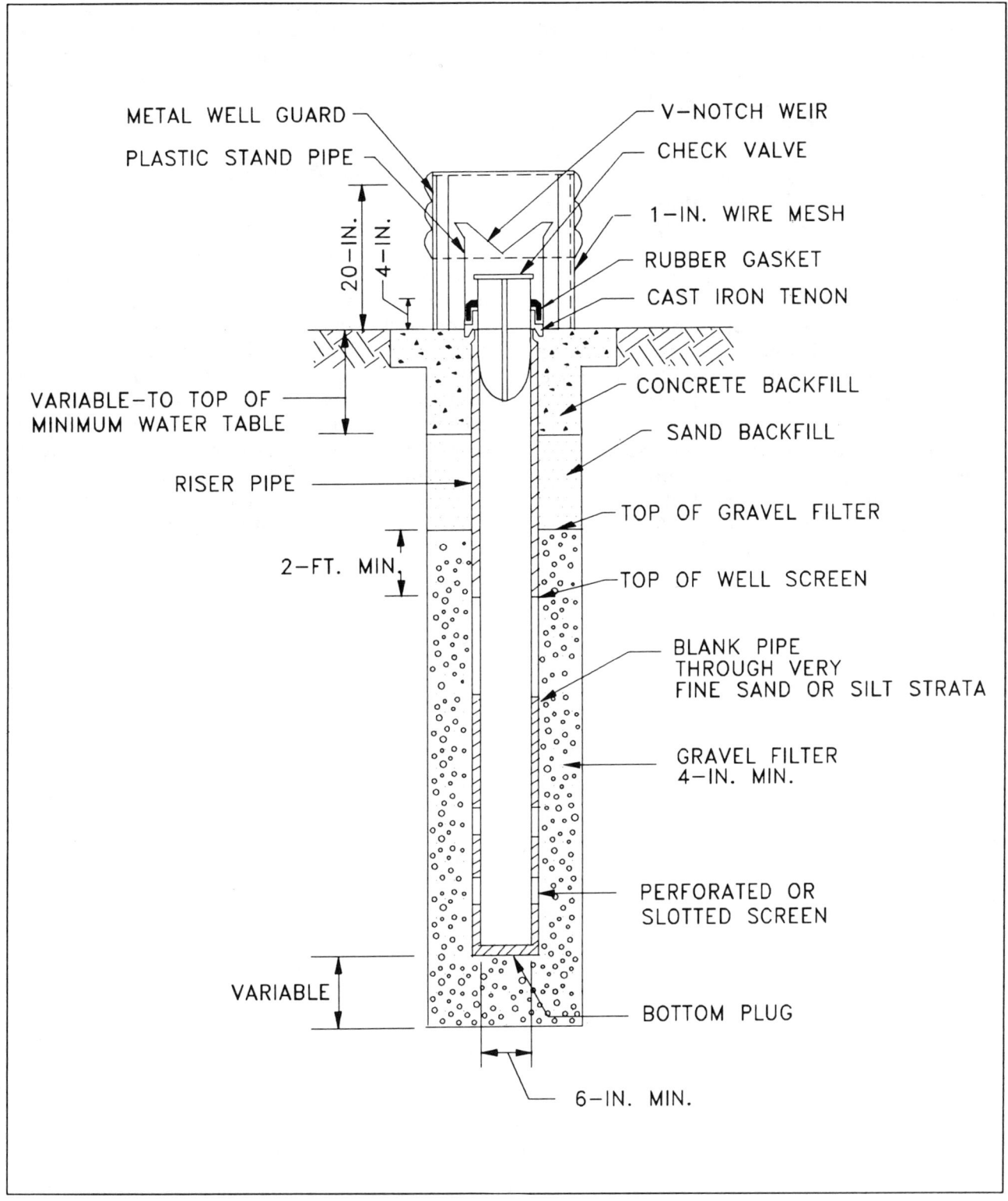

Figure 2-1. Typical relief well (after EM 1110-2-1913)

ing construction of Arkabutla Dam, Mississippi, completed in June 1943. The foundation consisted of approximately 30 ft of relatively impervious loess underlain by a pervious stratum of sand and gravel. The relief wells were installed to provide an added measure of safety with respect to uplift and piping along the downstream toe of the embankment. The relief wells consisted of 2-in. brass wellpoint screens 15 ft long attached to 2-in. galvanized wrought iron riser pipes spaced at 25-ft intervals located along a line 100 ft upstream of the downstream toe of the dam. The tops of the well screens were installed about 10 ft below the bottom of the impervious top stratum. The well efficiency decreased over a 12-year period by about 25 percent primarily as a result of clogging of the screens. However, the piezometric head along the downstream toe of the dam, including observations made at a time when the spillway was in operation, has not been more than 1 ft above the ground surface except at one location where a maximum excess head of 9 ft was observed (US Army Engineer Waterways Experiment Station 1958). Since these early installations, relief wells have been used at many levee locations to control excessive uplift pressures and piping through the foundation.

2-4. Other Applications

Pressure relief wells have also been used extensively beneath the stilling basins of spillways, outlet structures, and other hydraulic structures. In addition, wells have been employed to control excess hydrostatic pressures in outlet channels including areas immediately downstream of navigation locks. Often wells incorporated in structures have been located so that they discharge through collector pipes and manholes which are not readily accessible to cleaning and maintenance unless the structures are dewatered. An example of a relief well system incorporated into a toe drainage system for a dam is shown in Figure 2-2.

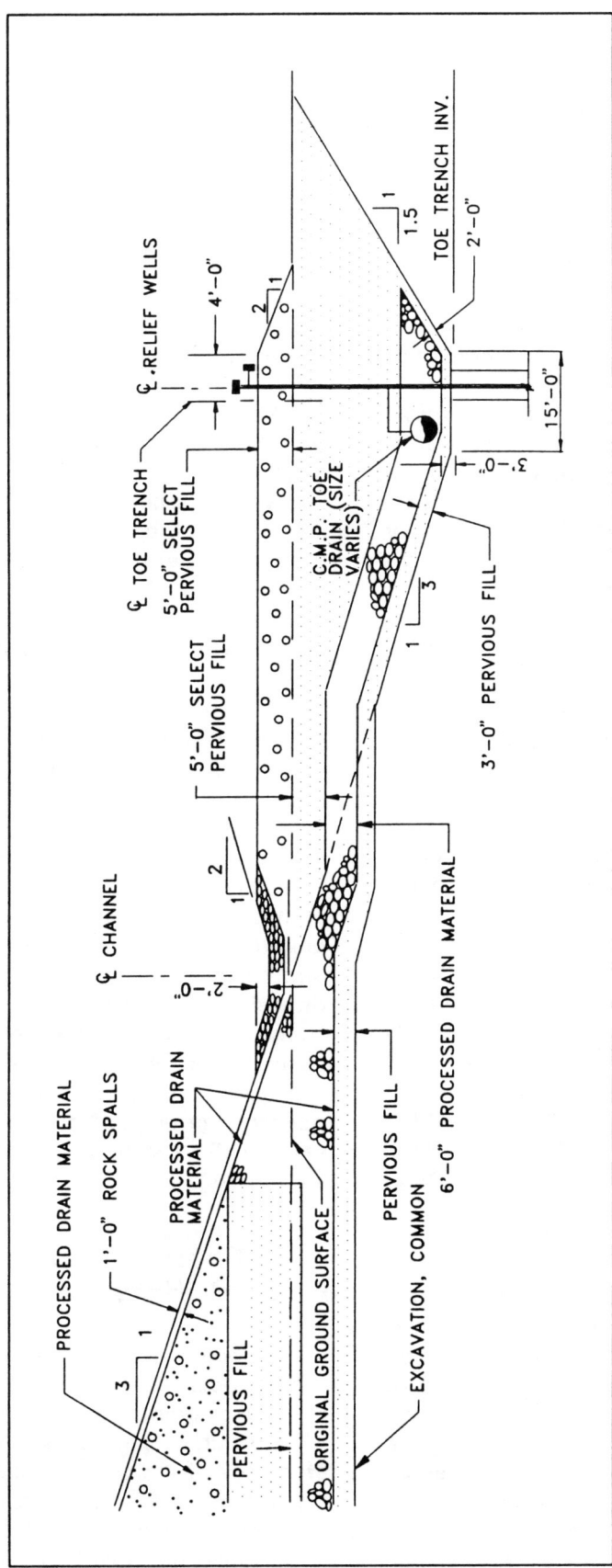

Figure 2-2. The drainage system with relief wells at Cochiti Dam

CHAPTER 3

BASIC CONSIDERATIONS

3-1. Foundation Investigations

The design of a relief well system should be preceded by thorough field and geologic studies conducted in accordance with EM 1110-1-1804. Sufficient borings should be made to define seepage entrance and exit conditions, the depth, thickness, and physical characteristics of the pervious strata, as well as the thickness and physical characteristics of the top stratum in upstream or riverside areas and downstream or landside areas. See Appendix B for further details. Particular attention should be given to the presence of buried channels and pervious abutments which could impact on underseepage estimates. An example of a generalized soil profile for relief well design along a levee reach is shown in Figure 3-1. The influence of surficial deposits on levee underseepage and on relief well design may be noted in Figure 3-2. High exit gradients and concentrations of seepage which may occur adjacent to clay-filled swales or channels will often govern the locations of individual relief wells. Where soil conditions vary along the proposed line of wells, the profile can be divided into a series of design reaches as shown in Figure 3-3. Additional borings, as subsequently described, should be made after completion of final design to ensure that a boring is located within 5 ft of each final well location. In general, samples should be taken at intervals not greater than 3 ft or at changes of soil strata, whichever occur first.

3-2. Foundation Permeability

Preliminary estimates of foundation permeability can be made from laboratory tests or correlations with grain size as described in EM 1110-2-1901. Because sampling operations do not necessarily indicate the relative perviousness of foundations containing large amounts of gravelly materials, field pumping tests are recommended to verify the foundation permeability on all projects where the use of pressure relief wells is being considered. The test well should fully penetrate the pervious aquifer, and a well flow meter should be used to determine the variations in horizontal permeability with depth. An example of data derived from a field pumping test conducted in this manner is shown in Figure 3-4. Field pumping test procedures for steady state and transient flow conditions are given in Appendix III to TM 5-818-5. Additional information, including procedures for field permeability tests in fractured rock, is given in EM 1110-2-1901. The vertical permeability of individual strata can be estimated from laboratory tests on undisturbed samples or determined from field pumping tests (Mansur and Dietrich 1965).

3-3. Anisotropic Conditions

Analytical methods for computing seepage through a permeable deposit are based on the assumption that the permeability of the deposit is isotropic. However, natural soil deposits are stratified to some degree, and the average permeability parallel to the planes of stratification is greater than the permeability perpendicular to these planes. Thus, the soil deposit actually possesses anisotropic permeability. To make a mathematical analysis of the seepage through an anisotropic deposit, the dimensions of the deposit must be transformed so that the permeability is isotropic. Each permeable stratum of the deposit must be separately transformed into isotropic conditions. In general, the simplest procedure is to transform the vertical dimensions with the horizontal dimensions unchanged.

3-4. Chemical Composition of Ground Waters

Some ground waters are highly corrosive with respect to elements of a pressure relief well or may contain dissolved minerals or carbonates which could in time cause clogging and reduced efficiency of the well. The chemical composition of the ground water, including river or reservoir supply waters, should be determined as part of the design investigation. Sampling, sample preservation, and chemical analyses of ground water is covered in handbooks (Moser and Huibregtse

BASIC CONSIDERATIONS

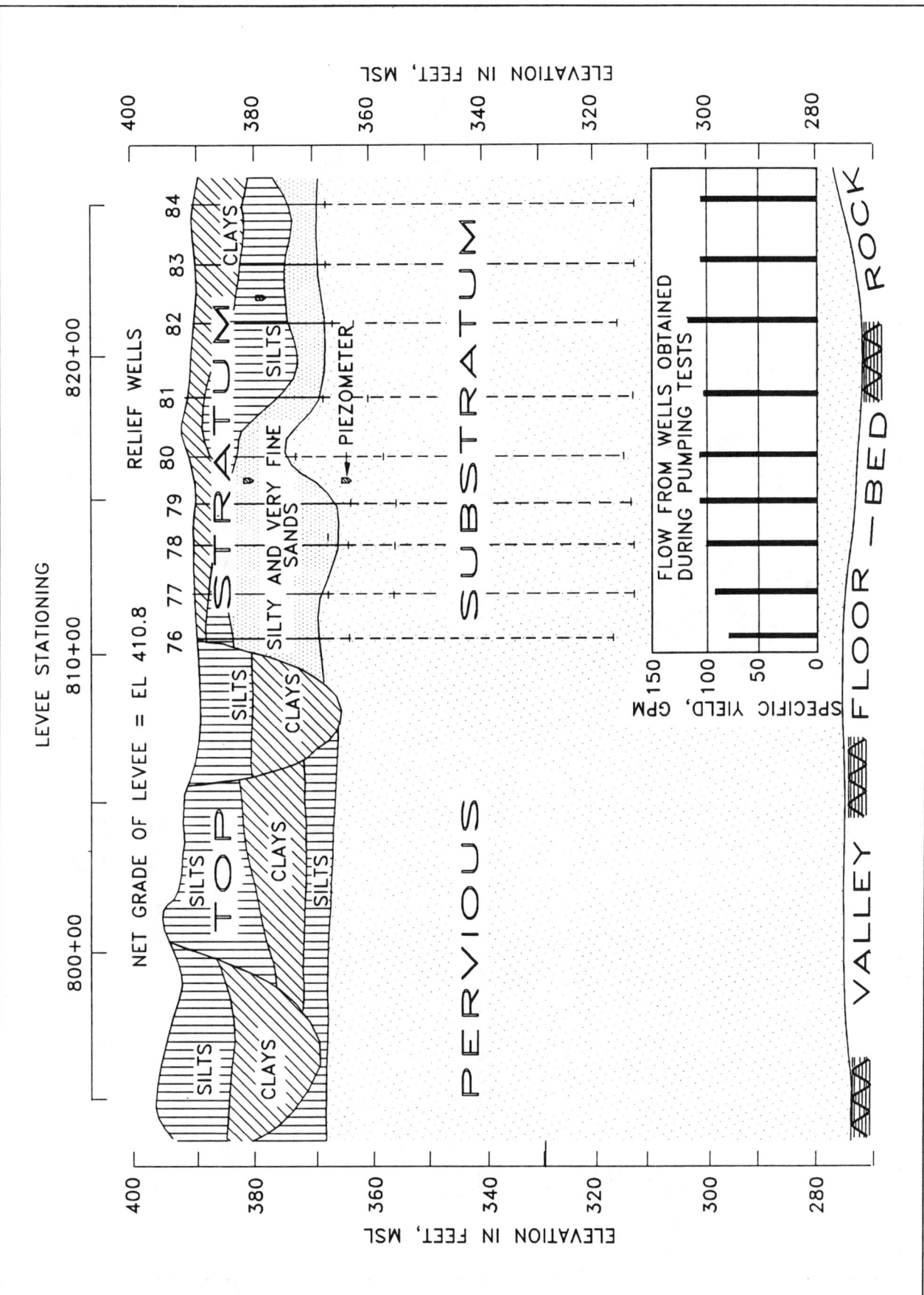

Figure 3-1. Soil profile, relief well, and piezometer installation data

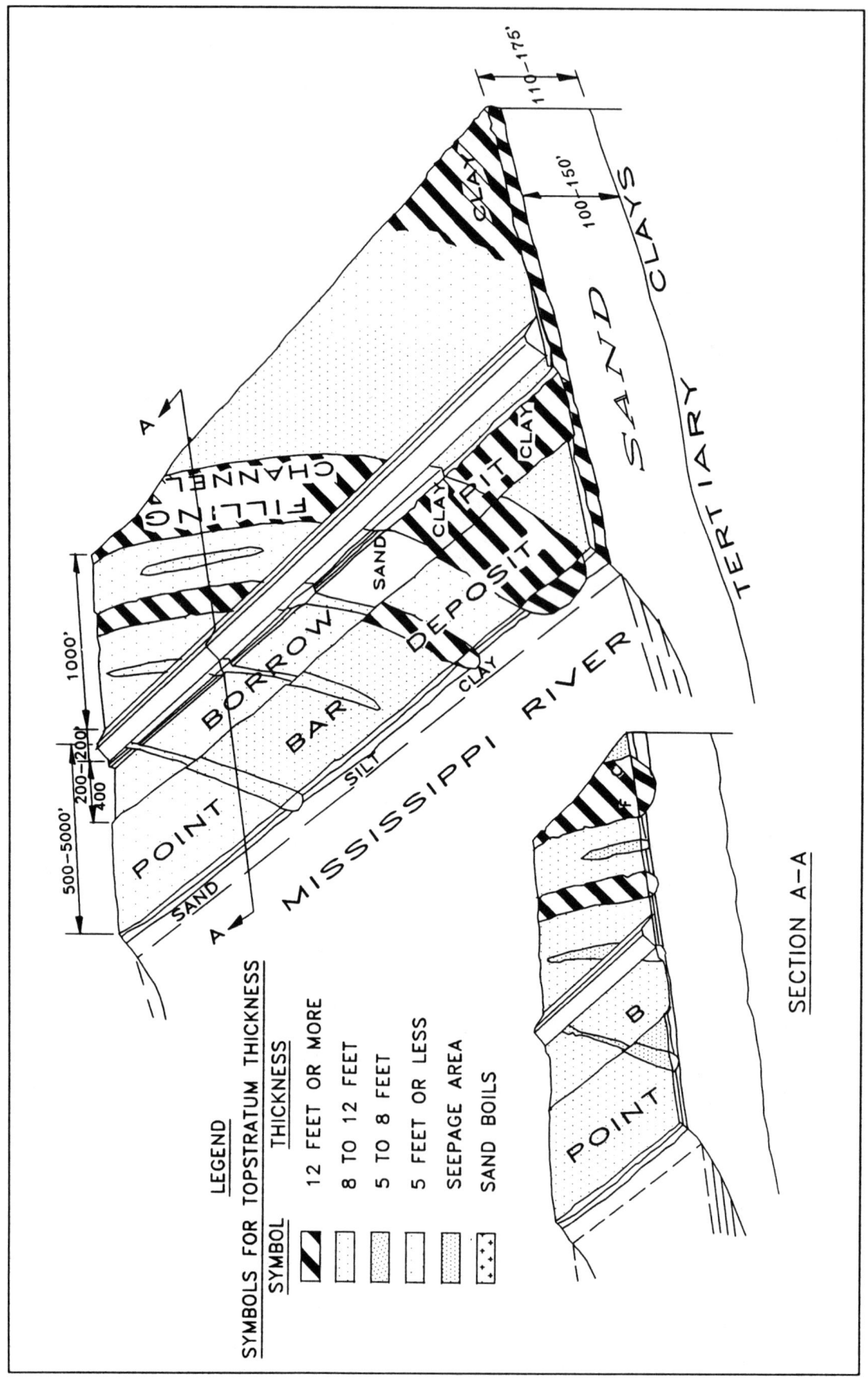

Figure 3-2. Seepage through point bar deposits

BASIC CONSIDERATIONS

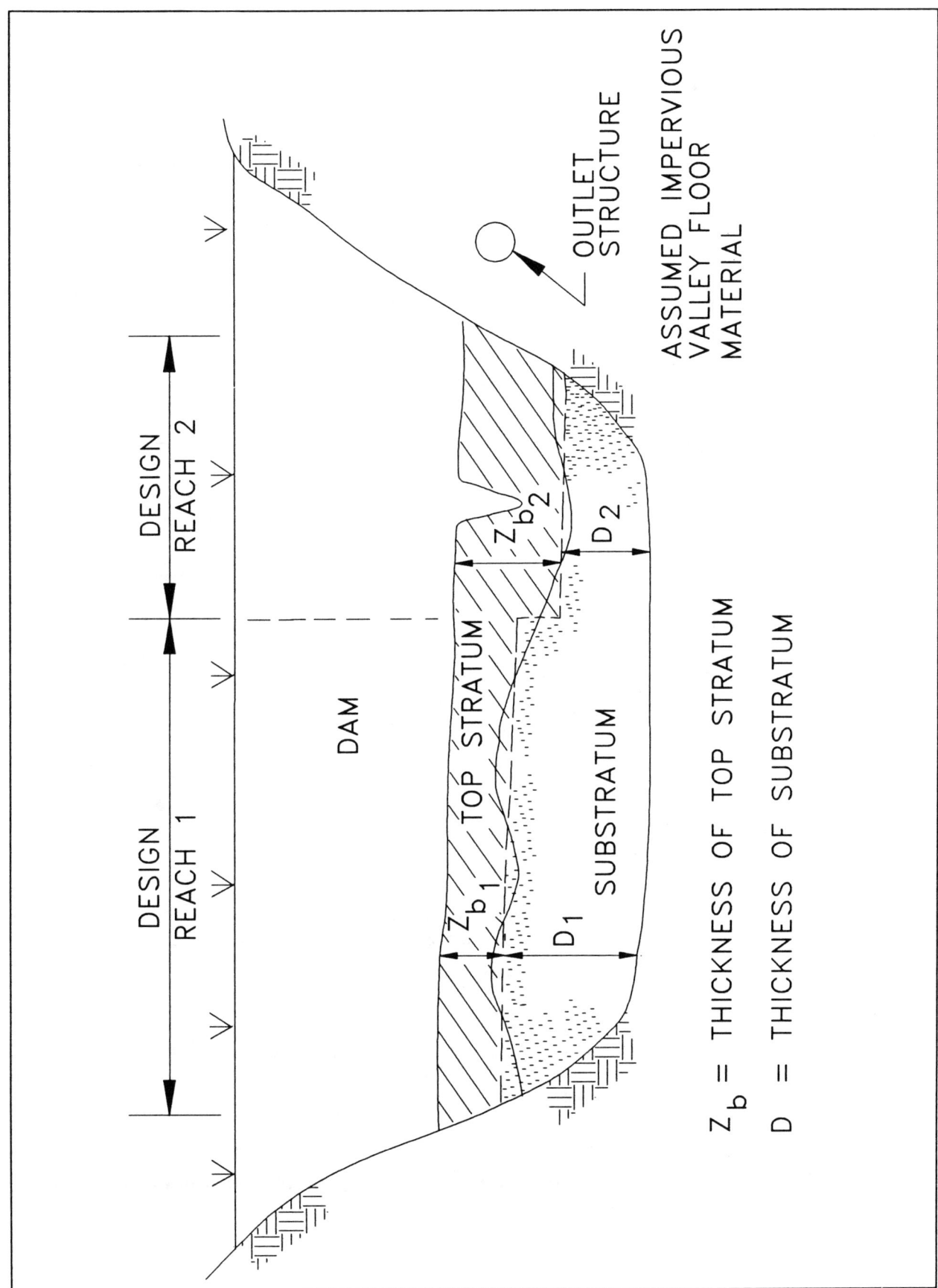

Figure 3-3. Profile of typical design reaches for relief well analysis

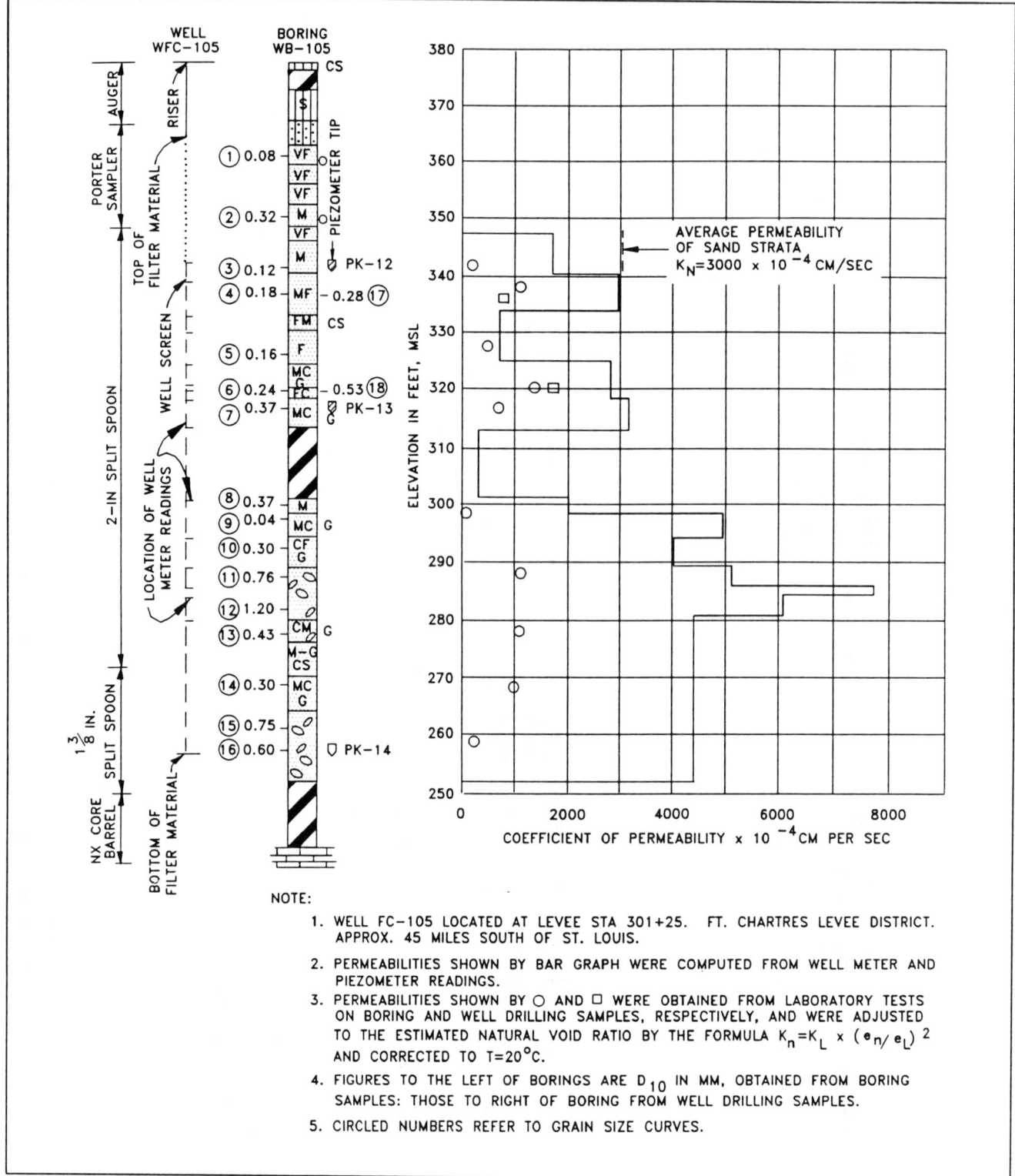

Figure 3-4. Coefficient of permeability and effective grain size of individual sand strata — Well FC-105

BASIC CONSIDERATIONS

Table 3-1. Indicators of Corrosive and Incrusting Waters[a]

Indicators of Corrosive Water	Indicators of Incrusting Water
1. A pH less than 7	1. A pH greater than 7
2. Dissolved oxygen in excess of 2 ppm[b]	2. Total iron (Fe) in excess of 2 ppm
3. Hydrogen sulfide (H_2S) in excess of 1 ppm detected by a rotten egg odor	3. Total manganese (MN) in excess of a 1 ppm in conjunction with a high pH and the presence of oxygen
4. Total dissolved solids in excess of 1,000 ppm indicates an ability to conduct electric current great enough to cause serious electrolytic corrosion	4. Total carbonate hardness in excess of 300 ppm
5. Carbon dioxide (CO_2) in excess of 50 ppm	
6. Chlorides (CL) in excess of 500 ppm	

Notes:
a. From TM 5-818-5.
b. ppm=parts per million.

1976, Environmental Protection Agency 1976). Indications of corrosive and incrusting waters are given in Table 3-1. The chemical composition of ground water is a major factor in the chemical and biological contamination of well screens and filter packs as described in Chapter 11.

3-5. Seepage Analysis

The determination of whether relief wells are needed is based on a seepage analysis which also provides the conditions for design of the relief well system. The seepage analysis defines the entrance and exit conditions and provides an estimate of substratum pressures which may exist under project flood conditions. On completed structures where piezometric data are available, seepage analyses are required to permit extrapolation of the data to the project flood conditions. The mathematical analysis of underseepage and substratum pressures is contained in Appendix B.

3-6. Allowable Heads

Whenever a structure underlain by pervious deposits is subjected to a differential hydrostatic head, seepage enters the pervious strata, creating an artesian pressure beneath the structure and downstream areas which could result in piping or failure by heave of the downstream top stratum. Pressure relief wells are designed to prevent piping and provide an adequate factor of safety, FS, with respect to uplift or heave. For this purpose, reduce the net head beneath the top stratum in downstream areas to an allowable value, h_a. The equation for FS is

$$FS = \frac{i_o}{i_c} = \frac{\gamma'/\gamma_w}{h_a/Z_t} = \frac{\gamma' Z_t}{\gamma_w h_a} \quad (3-1)$$

where

i_c = critical upward hydraulic gradient, the ratio of the submerged weight of soil, γ', to the unit weight of water, γ_w

Z_t = transformed thickness of downstream top stratum (see Appendix B)

The factor of safety with respect to uplift or heave normally should be at least 1.5. In addition to providing a minimum factor of safety with respect to uplift or heave (Condition a), relief wells may also be designed to ensure that piezometric heads in downstream areas are below ground surface, thereby preventing upward seepage from emerging beneath the downstream top stratum (Condition b). The latter condition usually applies to dams where visible seepage in downstream areas is undesirable and can be prevented by installing the wells with outlets in ditches or collector pipes along the embankment toe. The two conditions are illustrated in Figure 3-5.

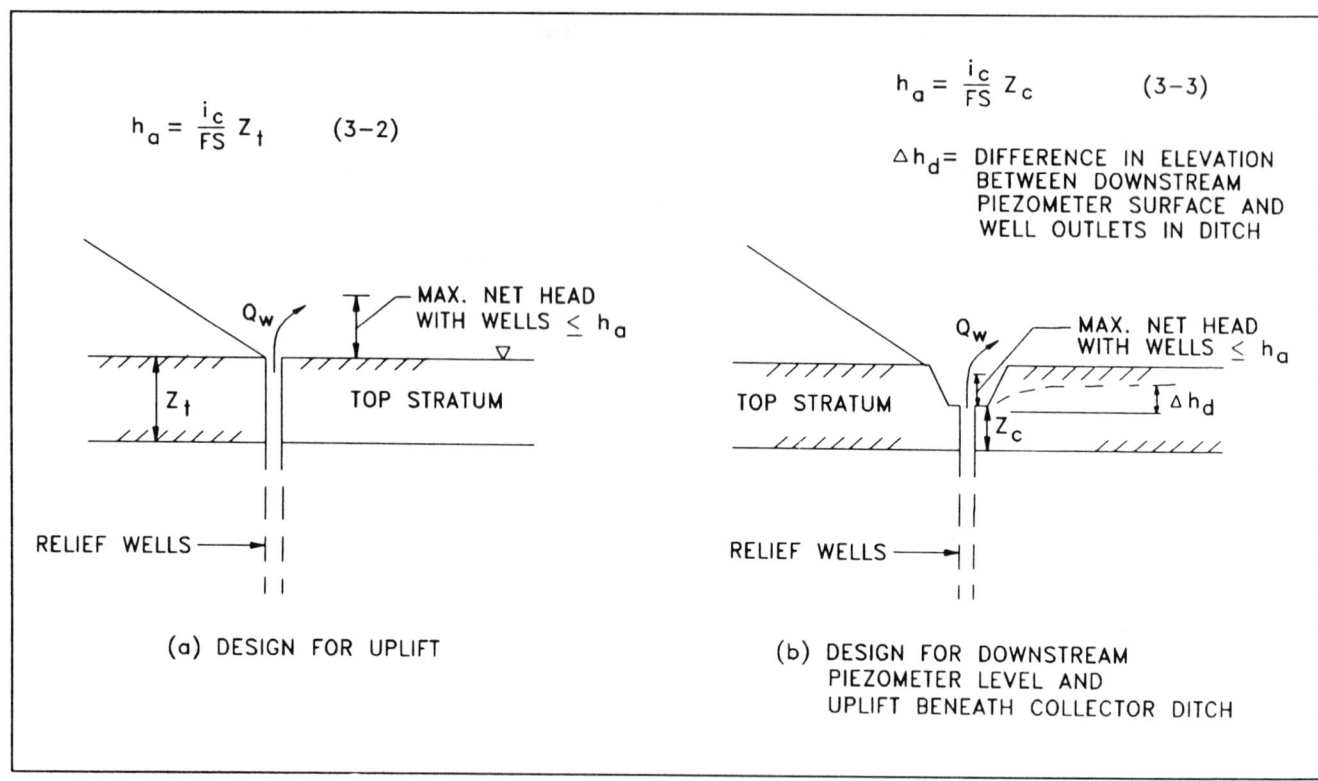

Figure 3-5. Determination of allowable heads in downstream toe area

A. CONDITION A. The allowable net head (h_a) under the top stratum of the downstream toe for this condition is given by

$$h_a = \frac{i_c}{FS} Z_t \qquad (3\text{-}2)$$

B. CONDITION B. The maximum downstream piezometric surface is defined by Δh_d which is the difference between this surface and the elevation of the well outlets corrected for well losses as subsequently described. For wells discharging into a collector ditch, the factor of safety with respect to uplift below the bottom of the collector ditch should be at least 1.5. The allowable net head under the top stratum below the bottom of the collector ditch for this condition is given by the equation

$$h_a = \frac{i_c}{FS} Z_c \qquad (3\text{-}3)$$

where Z_c is the transformed thickness of the downstream top stratum below the bottom of the collector ditch.

CHAPTER 4

ANALYSIS OF SINGLE WELLS

4-1. Assumptions

Analytical procedures for determining well flows and head distributions adjacent to single artesian relief wells are presented below. By definition, relief wells signify artesian conditions, and equations for artesian flow are applicable. In cases where wells are pumped, and gravity flow conditions exist, procedures for well analysis can be found in TM 5-818-5. It is assumed in the following analyses that all seepage flow is laminar or viscous, i.e., Darcy's Law is applicable. It is also assumed that steady state conditions prevail; the rate of seepage and rate of head reduction have reached equilibrium and are not time dependent. Unless otherwise indicated, the well is assumed to penetrate the full thickness of the aquifer.

4-2. Circular Source

Certain geologic or terrain conditions may require the assumption of a circular source of seepage. The formulas for a fully penetrating well located at the center of a circular source (see Figure 4-1) are

$$h_p = H - \frac{Q_w}{2\pi kD} \ln \frac{R}{r} \quad (4\text{-}1)$$

$$h_w = H - \frac{Q_w}{2\pi kD} \ln \frac{R}{r_w} \quad (4\text{-}2)$$

where

- h_p = head at point p between the well and the source
- H = head at the source
- Q_w = well discharge
- k, (k_f) = coefficient of permeability of pervious substratum
- D = thickness of pervious foundation
- R = radius of circular source (radius of influence)
- h_w = head at well
- r_w = radius of well

4-3. Noncircular Source

If geologic or terrain conditions indicate a noncircular source of seepage, the radius of influence, R, may be replaced by A_c, defined as an effective average of the distance from the well center to the external boundary. For a rectangular boundary of sides $2a$ and $2b$, the value of A_c is

$$A_c = \sqrt{\frac{4ab}{\pi}} \quad (4\text{-}3)$$

4-4. Infinite Line Source

Conditions may arise where the flow to the well originates from the bank of a river or canal reservoir or another body of water. In such cases, the bank or shoreline may act as an infinite line source of seepage. If leakage occurs through the top stratum, the effective distance to the infinite line source of seepage should be computed as discussed in Appendix B. The solutions for a single well adjacent to an infinite line source (see Figure 4-1) is determined using the method of images described by Muskat (1937), Todd (1980), and EM 1110-2-1901. The formulas are

$$h_p = H - \frac{Q_w}{2\pi kD} \ln \frac{r'}{r} \quad (4\text{-}4)$$

$$h_w = H - \frac{Q_w}{2\pi kD} \ln \frac{2S}{r_w} \quad (4\text{-}5)$$

where

r' = distance from point p to image well
r = distance from point p to real well
S = distance from real well to line source

A solution for h_p is also presented in terms of x and y coordinates in Figure 4-1 (Equation 4-6).

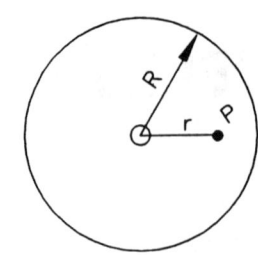

$$h_p = H - \frac{Q_w}{2\pi kD} \ln \frac{R}{r} \qquad (4-1)$$

$$h_w = H - \frac{Q_w}{2\pi kD} \ln \frac{R}{r_w} \qquad (4-2)$$

<u>CIRCULAR SOURCE</u>

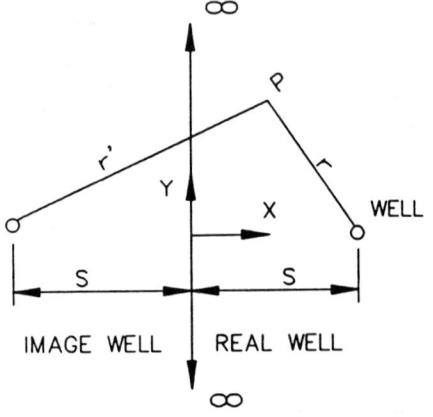

$$h_p = H - \frac{Q_w}{2\pi kD} \ln \frac{r'}{r} \qquad (4-4)$$

$$h_w = H - \frac{Q_w}{2\pi kD} \ln \frac{2S}{r_w} \qquad (4-5)$$

IN TERMS OF X AND Y COORDINATES

$$h_p = H - \frac{Q_w}{2\pi kD} \ln \left[\frac{y^2 + (x+S)^2}{y^2 + (x-S)^2}\right]^{\frac{1}{2}} \qquad (4-6)$$

<u>INFINITE LINE SOURCE</u>

$$h_w = H - \frac{Q_w}{2\pi kD} \ln \frac{4S}{r_w} \left[\frac{(c^2 - r_o^2)^2 + 4S^2 c^2}{c^2 - r_o^2 + \sqrt{(c^2 - r_o^2)^2 + 4S^2 c^2}}\right] \qquad (4-7)$$

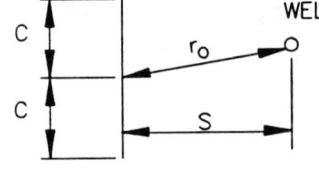

FOR WELL ON PERPENDICULAR BISECTOR, $r_o = S$

$$h_w = H - \frac{Q_w}{2\pi kD} \ln \frac{2S}{r_w}\left(1 + \frac{S^2}{c^2}\right) \qquad (4-8)$$

<u>FINITE LINE SOURCE</u>

Figure 4-1. Summary of equations for artesian flow to single well

4-5. Finite Line Source

In cases where the length of the source of seepage is relatively small compared to its distance from the well, the source may be considered as a finite line source. The solution for a single well adjacent to a finite line source was developed by Muskat (1937). The formulas, which are available only in terms of head at the well, are shown in Figure 4-1 (Equations 4-7 and 4-8).

4-6. Infinite Line Source and Infinite Line Sink

As discussed in Appendix B, a semipervious landside blanket can be replaced by a totally impervious top stratum and a theoretical line sink at an appropriate equivalent distance from the well. The theoretical line sink, parallel to the infinite line source, is referred to as an infinite line sink. A solution, based also on the method of images, considering one of the finite line sources as a sink, was developed by Barron (1948) and is shown in Figure 4-2.

4-7. Infinite Line Source and Infinite Barrier

The method of images is an extremely powerful tool for developing solutions to wells for various boundary conditions. Solutions for various boundary conditions including barriers are presented by Ferris, Knowles, Brown and Stellman (1962), Freeze and Cherry (1976), and Todd (1980). For example, a typical problem would be to calculate the discharge or heads for a single artesian well located between a river denoted by an infinite line source and a barrier such as a buried channel or rock bluff. In this case, the image well for the river would have a second image well with respect to the rock bluff which in turn would have an image with respect to the river and so on. A similar progression of image wells would be needed for the impermeable barrier (see EM 1110-2-1901). The image wells extend to infinity; however in practice, it is only necessary to include pairs of image wells closest to the real well because others have a negligible influence on the drawdown. A solution for this case was presented by Barron (1982) and is shown in Figure 4-3.

4-8. Complex Boundary Conditions

Oftentimes, geologic factors impose conditions which are difficult to simulate using circular or line sources and barriers. In such cases, flow net analyses or electrical analogy tests may be used to advantage especially when the aquifer thickness is irregular and three-dimensional analyses are required. The use of flow nets for the design of well systems is described by Mansur and Kaufman (1962). Methods for conducting three-dimensional electrical analogy tests are described by Duncan (1963), Banks (1965), and McAnear and Trahan (1972).

4-9. Partially Penetrating Wells

The previous equations are based on the assumption that the well fully penetrates the aquifer. For practical reasons, it is often necessary to use wells which only partially penetrate the aquifer. The ratio of flow from a partially penetrating artesian well to that for a fully penetrating well at the same drawdown is

$$\frac{Q_{wp}}{Q_w} = G_p \qquad (4-13)$$

or

$$Q_{wp} = G_p Q_w = \frac{2\pi k D (H - h_w) G_p}{\ln \frac{R}{r_w}} \qquad (4-14)$$

where

Q_{wp} = flow from partially penetrating well
G_p = flow correction factor for partially penetrating well

An approximate value of G_p can be obtained from the following equation developed by Kozeny (1933):

$$G_p = \frac{W}{D}\left(1 + 7\sqrt{\frac{r_w}{2w}} \cos \frac{\pi w}{2D}\right) \qquad (4-15)$$

where W/D is well penetration expressed as a decimal. An alternate equation developed by

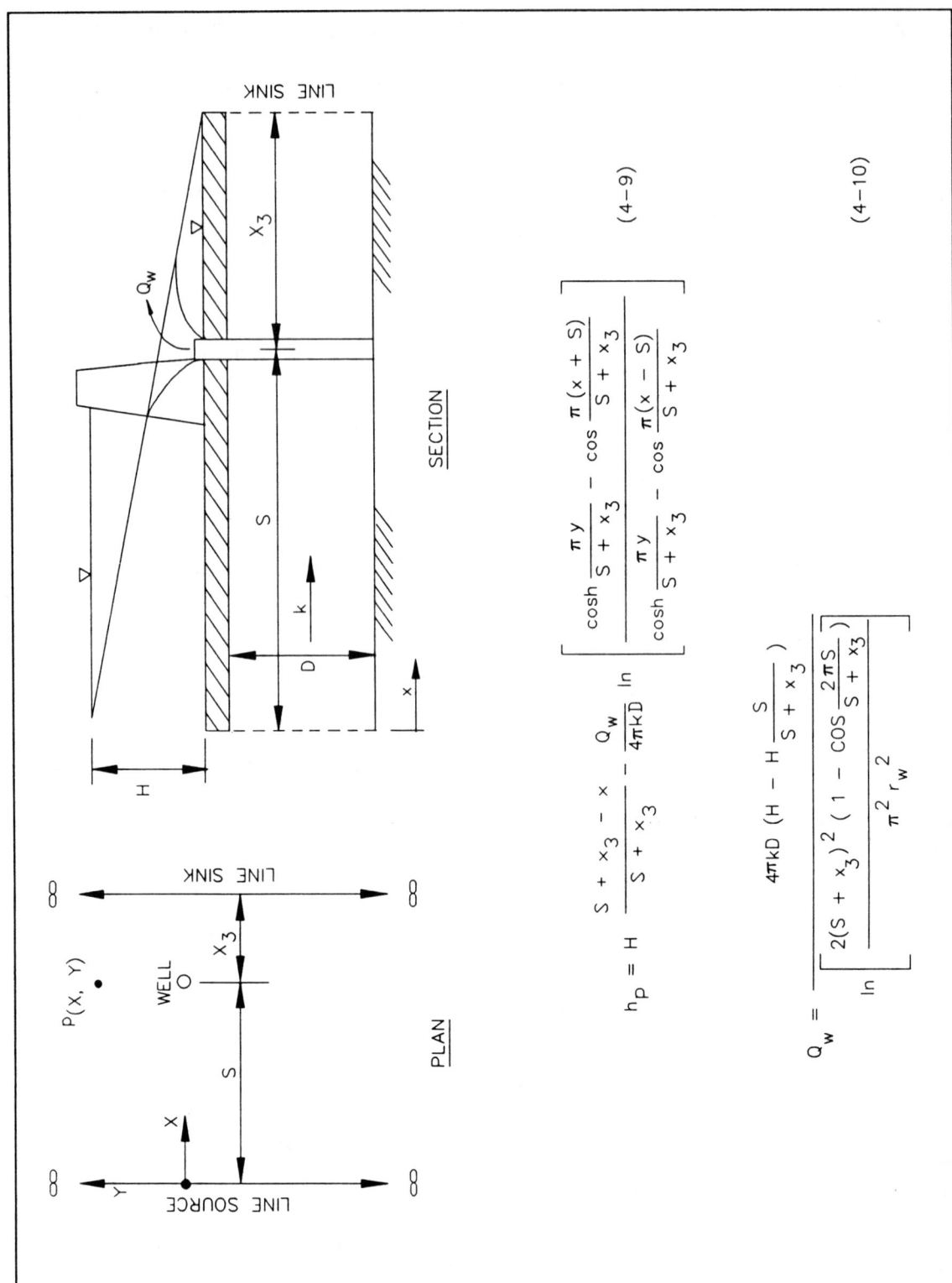

Figure 4-2. Drawdown for well between infinite line source and downstream sink

ANALYSIS OF SINGLE WELLS

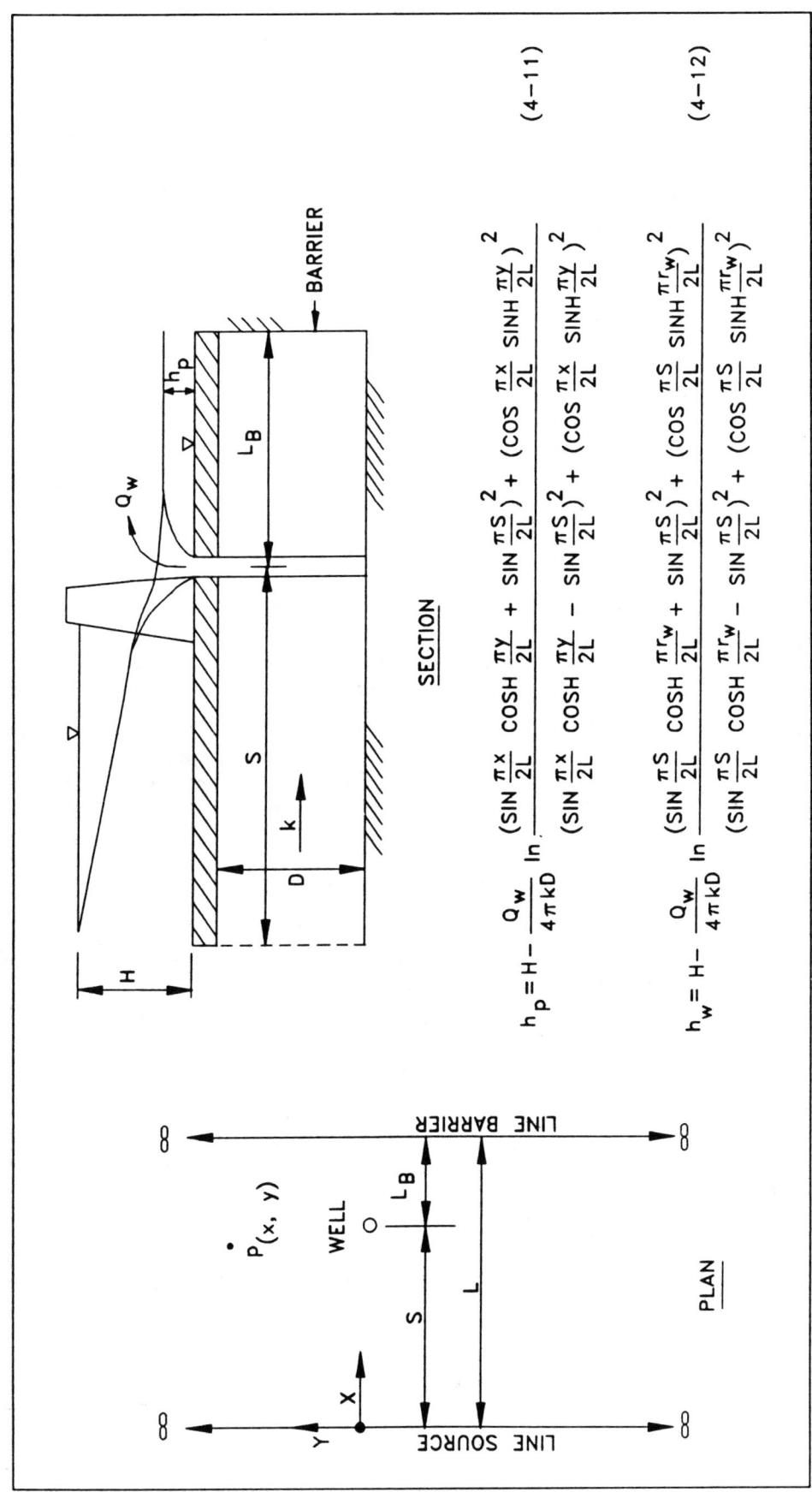

Figure 4-3. Drawdown for well between infinite line source and infinite barrier

Muskat (1937) assuming a constant flow per unit length of well screen is:

$$G_p = \frac{\ln \dfrac{R}{r_w}}{\dfrac{D}{2w}\left[2\ln\dfrac{4D}{r_w} - G(\bar{T})\right] - \ln\dfrac{4D}{R}} \quad (4\text{-}16)$$

where $G(\bar{T})$ is a function of W/D and approximate values from Harr (1962) are given in Table 4-1.

Values of G_p based on the above values for a typical well ($r_w = 1.0$ ft) with a radius of 1,000 ft are plotted in Figure 4-4. An empirical method for calculating the head of any point for partially penetrating wells is described by Warriner and Banks (1977). Limitations of empirical formulas for determining flows from partially penetrating wells are discussed in TM 5-818-5.

4-10. Effective Well Penetration

In a stratified aquifer, the effective well penetration usually differs from that computed from the ratio of the length of well screen to total thickness of aquifer. To determine the required length of well screen W to achieve an effective penetration $\overline{W}$ in a stratified aquifer, the procedure shown in Figure 4-5 can be used. It is assumed that the individual strata are anisotropic and each stratum is transformed into an isotropic stratum in accordance with the following equation:

$$\bar{d} = d\sqrt{\frac{k_h}{k_v}} \quad (4\text{-}17)$$

where

$\bar{d}$ = transformed vertical dimension
d = actual vertical dimension
k_h = permeability in the horizontal direction
k_v = permeability in the vertical direction

The horizontal dimension of the problem would remain unchanged in this transformation. The permeability of the transformed stratum to be used in all equations for flow or drawdown is as follows:

$$\bar{k} = \sqrt{k_h k_v} \quad (4\text{-}18)$$

where $\bar{k}$ is the transformed coefficient of permeability.

Table 4-1. Partially Penetrating Well Function, $G(\bar{T})$

W/D	$G(\bar{T})$
0.1	6.4
0.2	5.0
0.3	4.3
0.4	3.5
0.5	2.9
0.6	2.4
0.7	1.9
0.8	1.3
0.9	0.7
1.0	0.0

ANALYSIS OF SINGLE WELLS

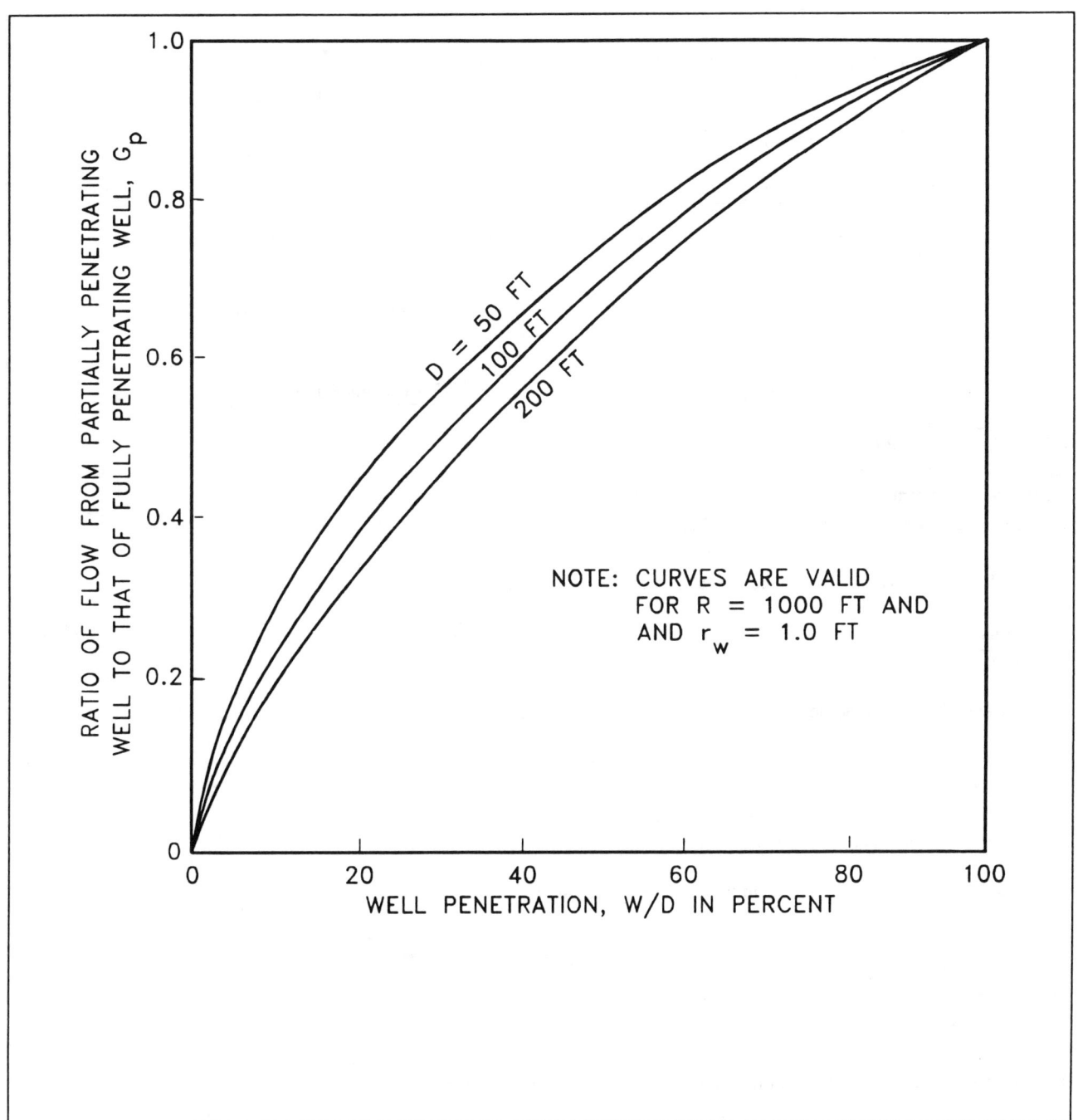

Figure 4-4. Flow to partially penetrating well with circular source

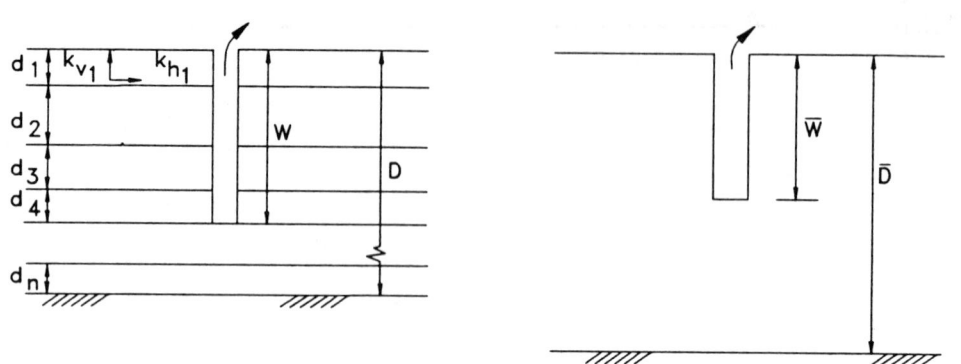

ACTUAL SECTION   TRANSFORMED SECTION

Actual well penetration = W
Effective well penetration = $\overline{W}$
Actual well penetration in percent = $W/D \times 100$
Effective well penetration in percent = $\overline{W}/\overline{D} \times 100$

1. Transform each layer into an isotropic layer of thickness $\overline{d}$ and permeability $\overline{k}$

$$\overline{d} = d\sqrt{\frac{k_h}{k_v}} \qquad (4\text{-}17) \qquad\qquad \overline{k} = \sqrt{k_h k_v} \qquad (4\text{-}18)$$

2. Calculate thickness of the equivalent homogeneous, isotropic acquifer, $\overline{D}$

$$\overline{D} = \sqrt{\sum_{m=1}^{m=n} d_m k_{Hm} \sum_{m=1}^{m=n} d_m/k_{vm}} \qquad (4\text{-}19)$$

n = number of strata, numbered from top to bottom

3. Calculate the effective permeability of the transformed aquifer, $\overline{k}_e$

$$\overline{k}_e = \sqrt{\frac{\sum\limits_{m=1}^{m=n} d_m k_{Hm}}{\sum\limits_{m=1}^{m=n} d_m/k_{vm}}} \qquad (4\text{-}20)$$

4. Calculate the effective well screen penetration into the transformed aquifer, $\overline{W}/\overline{D}$

$$\frac{\overline{W}}{\overline{D}} = \frac{\sum\limits_{o}^{W} \overline{d}\,\overline{k}}{\sum\limits_{m=1}^{m=n} \overline{d}_m \overline{k}_m} = \frac{\sum\limits_{o}^{W} \overline{d}\,\overline{k}}{\overline{D}\,\overline{k}_e} = \frac{\sum\limits_{o}^{W} dk_H}{\sum\limits_{m=1}^{m=n} dk_H} \qquad (4\text{-}21)$$

5. Determine actual well penetration required to achieve a given effective well penetration by successful trials

Figure 4-5. Determination of actual and effective well penetrations

CHAPTER 5

ANALYSIS OF MULTIPLE WELL SYSTEMS

5-1. General Equations

In most applications, a system of pressure relief wells in various arrays is required for the relief of substratum pressures or reduction of ground-water levels. In such cases, analyses must be made to determine the number and spacing of wells to meet these requirements. The head at any point p produced by a system of fully penetrating artesian wells was first determined by Forcheimer (1914). His general equation as later modified by Dachler (1936) is

$$h_p = H_1 - \frac{1}{2\pi kD}\left(Q_{w1} \ln \frac{R_1}{r_1} + Q_{w2} \ln \frac{R_2}{r_2} + \cdots Q_{wn} \ln \frac{R_n}{r_n}\right) \quad (5\text{-}1)$$

or

$$h_p = H_1 - \frac{1}{2\pi kD}\left(\sum_{i=1}^{i=n} Q_{wi} \ln \frac{R_i}{r_i}\right) \quad (5\text{-}2)$$

where

- H = gross head on system
- n = number of wells in group
- Q_{wi} = discharge from ith well
- R_i = radius of influence of ith well
- r_i = distance from ith well to point at which head is computed

The head h_{wj}, at any well, e.g. well j, in a system of n wells is determined from the equation

$$h_{wj} = H_1 - \frac{1}{2\pi kD}\left(Q_{wj} \ln \frac{R_j}{r_{wj}} + \sum_{i=1}^{i=n-1} Q_{wi} \ln \frac{R_i}{r_{i,j}}\right) \quad (5\text{-}3)$$

where

- Q_{wj} = flow from well j
- R_j = radius of influence of well j
- r_{wj} = effective well radius of well j
- r_{ij} = distance from each well to well j

The other symbols are as defined previously. Equations 5-1 and 5-3 as well as subsequent equations for multiple-well systems are based on the principle of superposition. Thus, the head at a given well in a system of wells is equal to that resulting from this well flowing as if no other wells were present minus the head reduction caused at the well due to flow from the remaining wells. In most applications, the radius of influence is large compared to the distance between wells and can be considered as constant. When wells are pumped as in a dewatering system, the values of Q_{wi} are known (or assumed). However, when n wells are used for pressure relief where they flow under artesian head conditions, the flow from each well must be computed taking into account the discharge elevation of each well. The procedure requires the solution of n simultaneous equations to determine individual well flows.

5-2. Empirical Method

An empirical method developed by Warriner and Banks (1977) using the results of electrical analogy studies by Duncan (1963) and Banks (1965) can be used to determine the head at any point within a random array of fully or partially penetrating wells. The method, described in EM 1110-2-1901, is also valid for arbitrarily-shaped source boundaries. A FORTRAN computer code is provided by Warriner and Banks (1977).

5-3. Circular Source

A. GENERAL CASE. The general equations for a group of fully penetrating wells subject to seepage from a circular source with radius R are shown in Figure 5-1. It is assumed that the radius R is large with respect to the distances between wells and that the flows from each well are equal. As indicated previously in the case of variable well discharges, the procedure requires the solution of n simultaneous to solve for individual well flows.

B. CIRCULAR ARRAY OF WELLS. A special case consists of a circular array of n wells equally spaced along the circumference of a circle of radius r_c, the center of which is also the center of a circular source of seepage of radius R. The general equations are shown in Figure 5-2.

C. OTHER WELL ARRAYS. For other multiple-well systems within a circular source, see Muskat (1937), Banks (1963), and TM 5-818-5.

5-4. Wells Adjacent to Infinite Line Source with Impervious Top Stratum

Where wells are located adjacent to a source which can be approximated as an infinite line source and the pervious stratum is overlain by an impervious top stratum extending landward to a great distance, a solution for heads and well flows is obtained using the method of images. The equations are shown in Figure 5-3 for the case of (a) equal well discharges and (b) variable well discharges. As noted previously, case (b) requires the solution of n simultaneous equations to determine individual well flows.

5-5. Infinite Line of Wells

An infinite line of wells refers to a system of wells that conforms approximately to the following idealized conditions:

A. The wells are equally spaced and identical in dimensions.

B. The pervious stratum is of uniform depth and permeability along the entire length of the system.

C. The effective source of flow and the effective landside exit or block, if present, are parallel to the line of the wells.

D. The boundaries at the ends of the system are impervious, normal to the line of the wells, and at a distance equal to one-half the well-spacing beyond the end of the well system. For the above conditions, the flow to each well and the pressure distribution around each well are uniform for all wells along the line. Therefore, there is no flow across planes centered between wells and normal to the line, hence no overall longitudinal component of the flow exists anywhere in the system. The term infinite is applied to such a system because it may be analyzed mathematically by considering an infinite number of wells; the actual number of wells in the system may be from one to infinity.

5-6. Top Stratum Conditions

The permeability and lateral extent of the top stratum landward of an infinite line of wells can have a pronounced effect on the performance of the well system. The assumption of a completely impervious top stratum extending landward to infinity is a convenient assumption for which theoretical solutions are available. However, this condition is rarely realized in practice. A more general condition occurs when the impervious top stratum extends landward a finite distance terminating at a line sink. This condition is also applicable with respect to results at the well line to the case of a semipervious top stratum which can be converted to an equivalent length of impervious top stratum using appropriate blanket formulas. The two conditions are illustrated in Figure 5-4 together with assumed head distributions with and without relief wells including the effects of well losses. Calculations of the corrected net head on the well system, h, should also take into consideration any extension of the well riser above tailwater elevation. A third condition occurs when the pervious substratum is blocked at some point landward of the well line. Theoretical solutions for the three conditions follow.

5-7. Infinite Line of Wells, Impervious Top Stratum

The head midway between wells and the well flows for the case of an impervious top stratum extending landward a great distance ($L_3 = \infty$) may be calculated using the method of multiple images (after Muskat 1937, Middlebrooks and Jervis 1947). Solutions are shown in Figure 5-5 for the case of no well losses. Equations 5-14 through 5-17 are applicable to both fully penetrating and partially penetrating wells. The latter make use of the so-called well factors, Θ_a and Θ_m.

ANALYSIS OF MULTIPLE WELL SYSTEMS

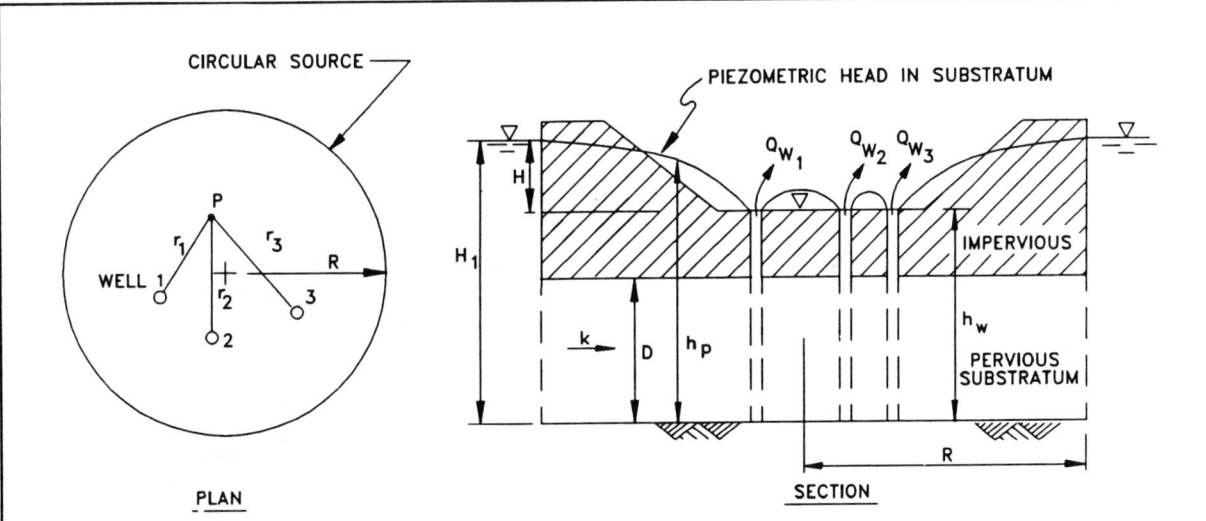

The head at Point P is:

$$h_p = H_1 - \frac{1}{2\pi kD}\left(Q_{w1} \ln \frac{R}{r_1} + Q_{w2} \ln \frac{R}{r_2} + \ldots Q_{wn} \ln \frac{R}{r_n}\right) \quad (5\text{-}4)$$

or

$$h_p = H_1 - \frac{1}{2\pi kD} \sum_{i=1}^{i=n} Q_{wi} \ln \frac{R}{r_i} \quad (5\text{-}5)$$

The head at Well 1 is:

$$h_{w1} = H_1 - \frac{1}{2\pi rD}\left(Q_{w1} \ln \frac{r}{r_{w1}} + \sum_{i=2}^{i=n} Q_{wi} \ln \frac{R}{r_i}\right) \quad (5\text{-}6)$$

If all wells have the same radius, and discharge at the same elevation, h_w, then the well discharges are equal and given by:

$$Q_w = \frac{2\pi kD(H_1 - h_w)}{n \sum_{n=1}^{n=n} \ln \frac{R}{r_w}} \quad (5\text{-}7)$$

Figure 5-1. Random array of fully penetrating wells with a circular source

5-8. Well Factors

The well factor, Θ_a, is the "extra length" or average uplift factor, and Θ_m is the midwell uplift factor. For fully penetrating wells,

$$\theta_a = \frac{1}{2\pi} \ln \frac{a}{2\pi r_w} \quad (5\text{-}18)$$

$$\theta_m = \frac{1}{2\pi} \ln \frac{a}{\pi r_w} \quad (5\text{-}19)$$

Approximate solutions for the well factors for various well penetrations were developed by Bennett and Barron (1957). More theoretically exact solutions were developed by Barron (1982) and verified by electrical analogy tests. The theoretical results are shown in Table 5-1 and plotted in Figures 5-6 and 5-7 together with the data from the electrical analogy tests. As there is a linear relation between the well factors and log a/r_w for values of a/r_w greater than about 20, the well factors are shown in terms of values at $a/r_w = 100$. The well factors at any other value of a/r_w are given by the following equations:

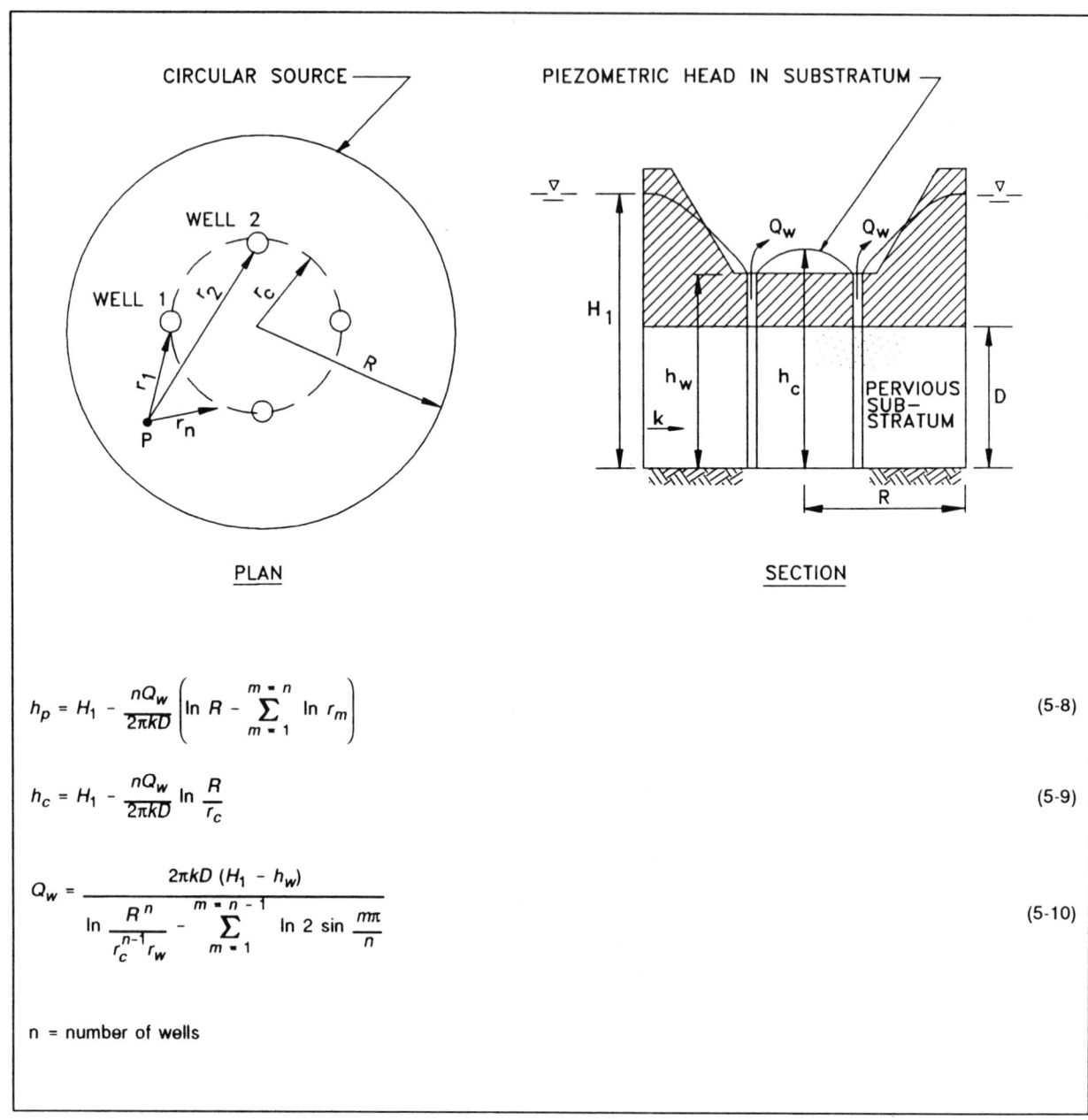

Figure 5-2. Circular array of fully penetrating artesian wells with a circular source

$$\theta_a = \theta_a(a/r_w = 100) + \Delta\theta(\log a/r_w - 2) \tag{5-20}$$

$$\theta_m = \theta_m(a/r_w = 100) + \Delta\theta(\log a/r_w - 2) \tag{5-21}$$

where $\Delta\Theta$ is obtained from Table 5-1. Values of the well factors may also be obtained from the nomograph from EM 1110-2-1901 shown in Figure 5-8 (after Bennett and Barron 1957). The nomograph though based on approximate solutions, reasonably accurate for well penetrations greater than 25 percent. A computer program for well design based on the Figure 5-8 was developed by Conroy (1984).

5-9. Infinite Line of Wells, Impervious Top Stratum of Finite Length

In many instances, the impervious top stratum landward of a line of wells is of finite length, and the boundary edge can be considered as a line sink. The presence of exposed borrow pits or

ANALYSIS OF MULTIPLE WELL SYSTEMS 25

Table 5-1. Theoretical Values of Q_a and Q_m

W/D	D/a	a/r_w	Θ_a	Θ_m	$\Delta\Theta$
100%	All values	100	0.440	0.550	1.00
75%	0.25	100	0.523	0.633	0.489
	0.50		0.563	0.667	
	1.0		0.606	0.681	
	2.0		0.678	0.682	
	3.0		0.748	0.682	
	4.0		0.818	0.682	
50%	0.25	100	0.742	0.851	0.733
	0.40		0.857	0.955	
	1.0		0.983	1.012	
	2.0		1.175	1.024	
	3.0		1.361	1.024	
	4.0		1.547	1.024	
25%	0.25	100	1.225	1.335	1.466
	0.50		1.569	1.622	
	1.0		1.926	1.908	
	2.0		2.390	2.024	
	3.0		2.798	2.047	
	4.0		3.199	2.075	
15%	0.25	100	1.662	1.772	2.077
	0.50		2.310	2.401	
	1.0		2.970	2.938	
	2.0		3.747	3.293	
	4.0		4.941	3.432	
10%	0.25	100	1.908	2.018	3.298
	0.50		2.934	3.025	
	1.0		3.977	3.941	
	2.0		5.139	4.649	
	4.0		6.814	5.071	
5%	0.25	100	1.778	1.887	6.963
	0.50		3.879	3.969	
	1.0		6.063	6.021	
	2.0		8.377	7.864	
	4.0		11.144	9.283	

other seepage exits landward of the well line can be simulated by a line sink. The head distribution beneath the top stratum without wells varies linearly from 100 percent of the net head at the effective source of seepage to 0 percent at the line sink. The conditions are illustrated in Figure 5-4 (b). These conditions are also applicable to the case of a semipervious landside blanket after conversion to an equivalent length of impervious blanket x_3. Equations for the head midway between wells and well flows are shown in Figure 5-9. The equations are applicable to both fully penetrating and partially penetrating well systems. The equations in Figure 5-9 apply to the case of no well losses. If well losses are considered, substitute h for H as shown in Figure 5-4 (b).

5-10. Infinite Line of Wells, Impervious Top Stratum Extending to Blocked Exit

Pervious foundations seldom extend landward to a great distance. Blockades generally occur because of the presence of old clay-filled channels or upland formations. If the distance from the line of wells is large, then the approximation of an infinite landward extent is reasonable. If the distance from the line of wells is less than the well spacing, then the error due to the approximation may be significant. The equations for the head midway between wells and well flow are shown in Figure 5-10 with exact equations for the case of

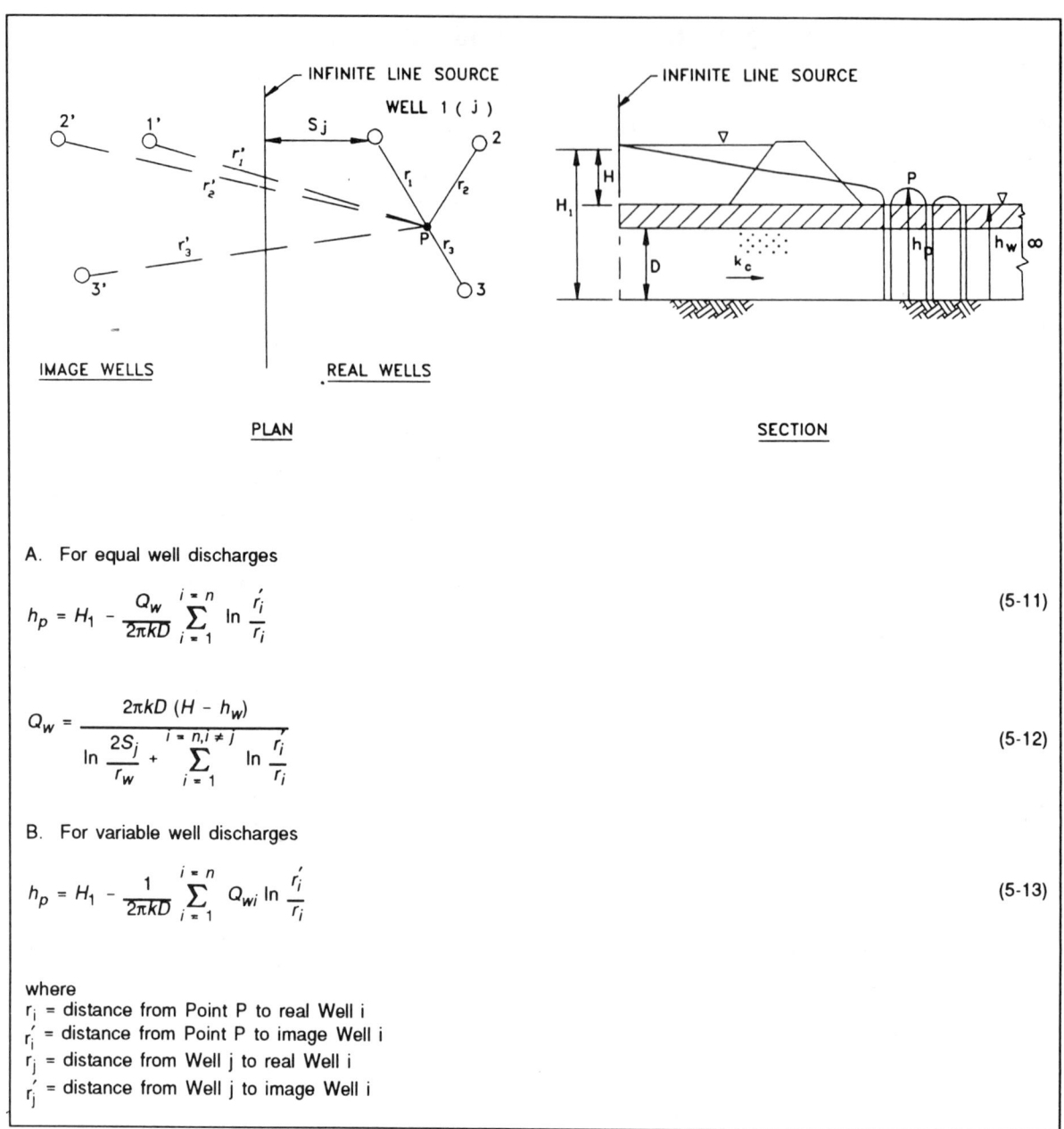

Figure 5-3. Multiple wells adjacent to infinite line source — general case

fully penetrating wells and reasonably accurate equations for both fully and partially penetrating wells where the distance to the blocked exit is greater than one-half times the well-spacing. The presence of a blocked exit can be ignored if the equivalent length of landside impervious top stratum is less than L_B.

5-11. Infinite Line of Wells, Discharge Below Ground Surface

In many well installations, the well outlets are located below the ground surface to prevent any seepage upward through the top stratum. Under this condition, the blanket formulas are inapplicable and the top stratum is assumed to be impervious. Solutions are obtained using equations in Figure 5-5, with h_d at or below ground surface, assuming $\Delta h_d = H_{av}$.

5-12. Infinite Line of Wells, No Top Stratum

A special case may exist in which there is no landside top stratum and wells are needed to lower the heads below the landward ground sur-

ANALYSIS OF MULTIPLE WELL SYSTEMS

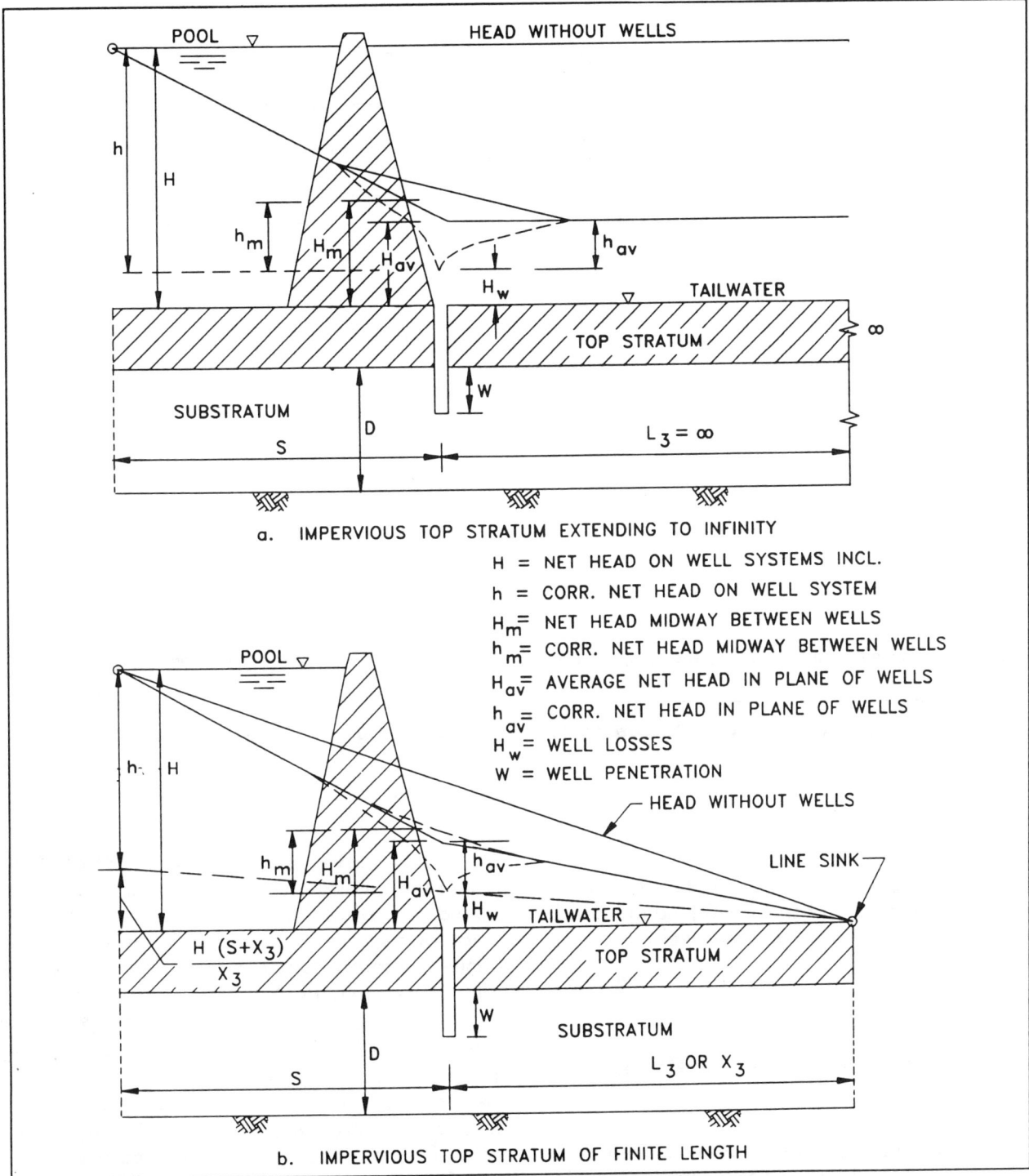

Figure 5-4. Infinite line of wells with infinite or finite impervious top stratum — general case

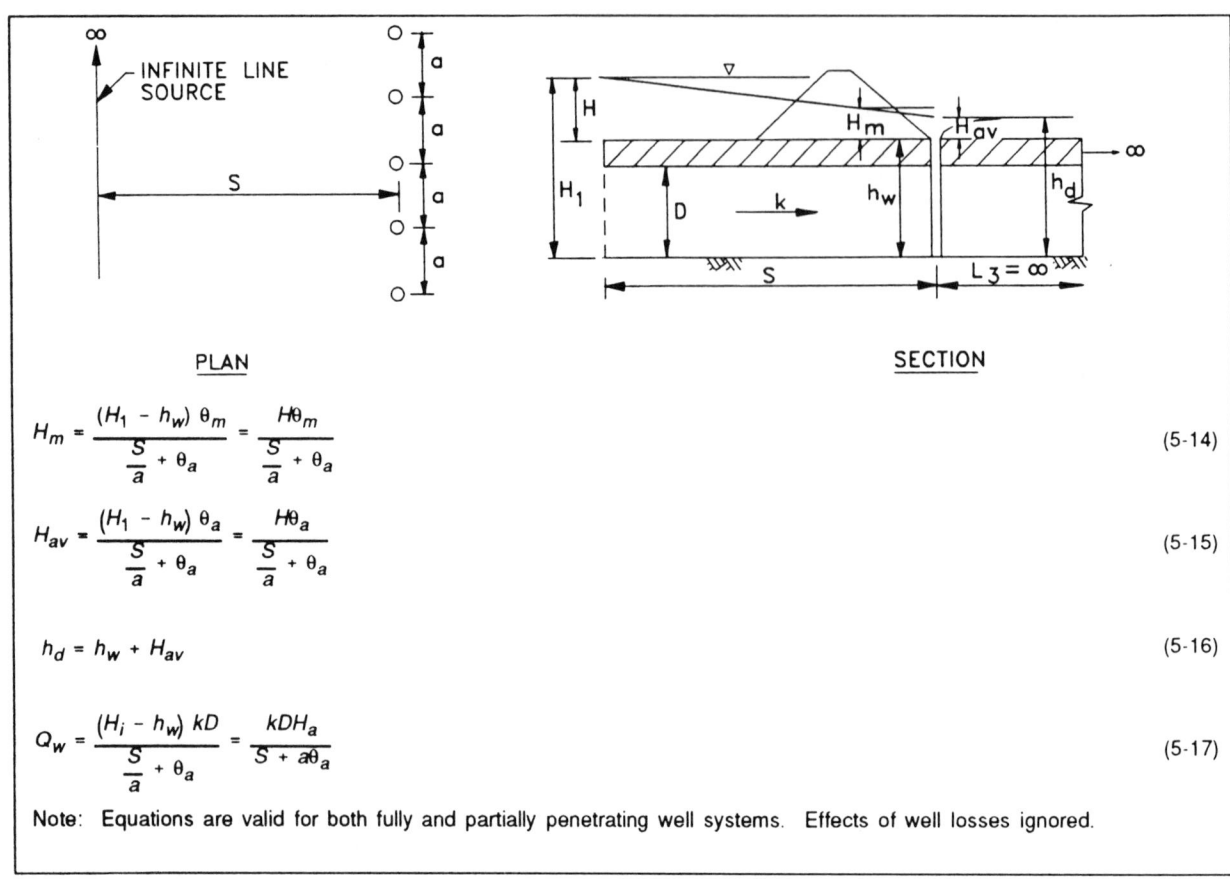

Figure 5-5. Infinite line of wells parallel to infinite line source — impervious top stratum

face. The flow in this case is a combination of artesian and gravity flow, and the equations shown in Figure 5-11 (Johnson 1947) may be used to estimate heads midway between wells and well flows for design.

5-13. Finite Well Lines, Infinite Line Source

The essential difference between finite and infinite well lines is the presence or absence of an appreciable component of flow parallel to the line of wells, resulting in nonuniform distribution of heads midway between wells and well discharges.

A. IMPERVIOUS TOP STRATUM. Where the landside top stratum is impervious and extends landward to infinity, the solution for a linear array of equispaced wells parallel to an infinite line source can be obtained using the equations shown in Figure 5-3.

B. IMPERVIOUS TOP STRATUM OF FINITE LENGTH. In the case of an impervious top stratum extending to a finite distance landward of the well line or in the case of a semipervious landside top stratum converted to an equivalent length of impervious top stratum, theoretical solutions for finite well lines are not available. Empirical solutions based on electrical analogy tests are presented in EM 1110-2-1905. The application of these solutions for design is discussed in Chapter 7.

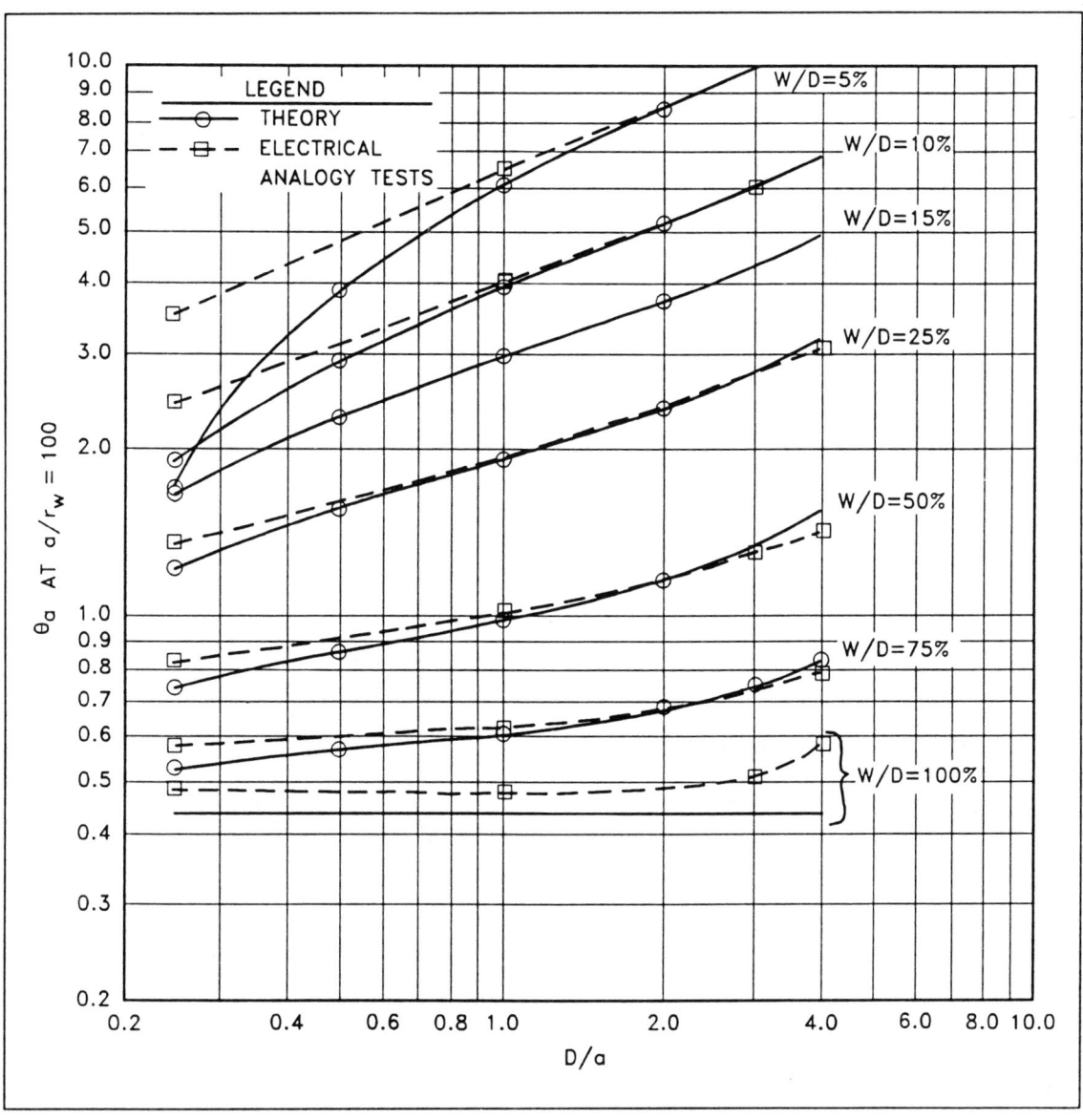

Figure 5-6. Theoretical values of average uplift factor (after Barron 1982)

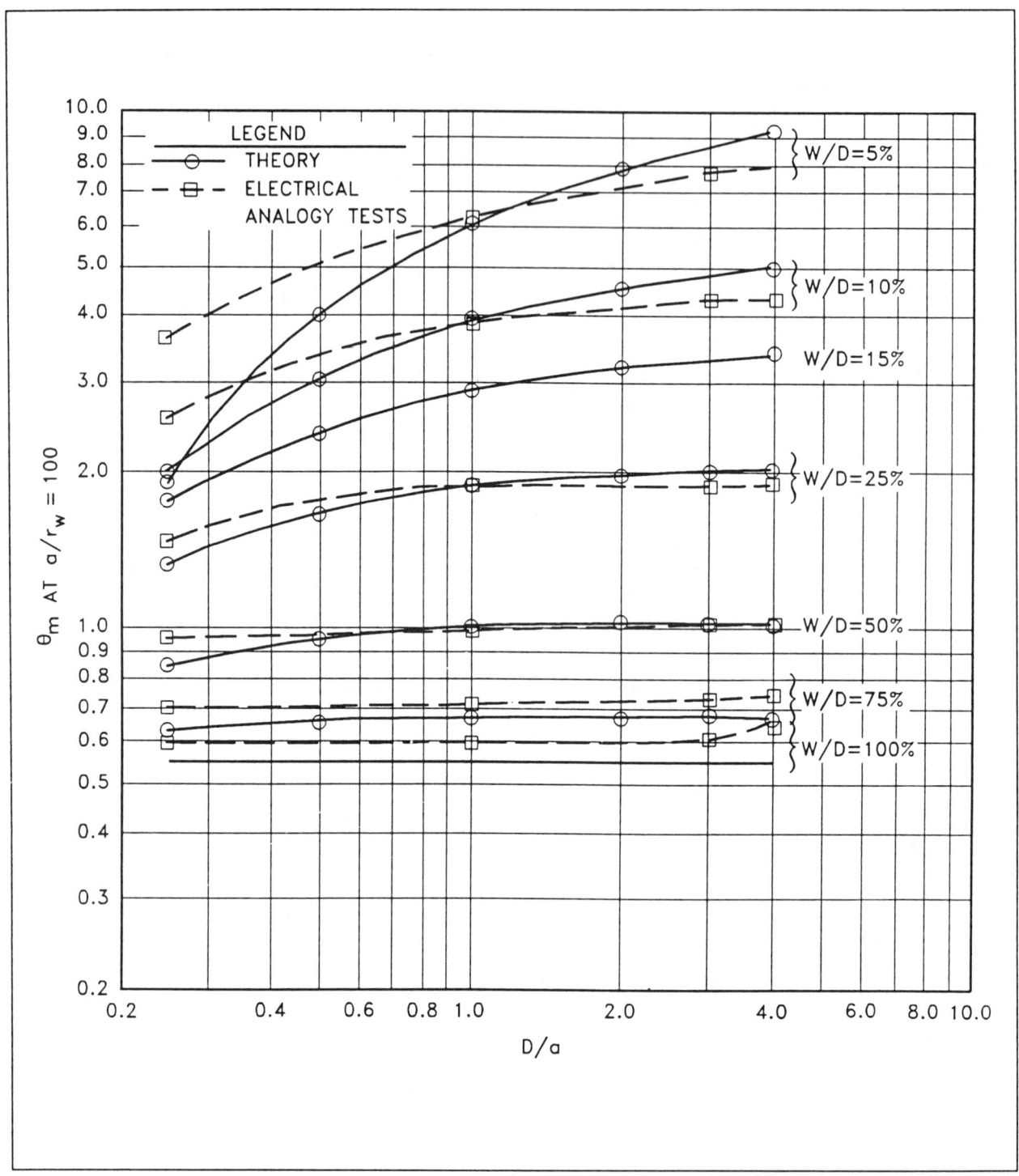

Figure 5-7. Theoretical values of midwell uplift factor (after Barron 1982)

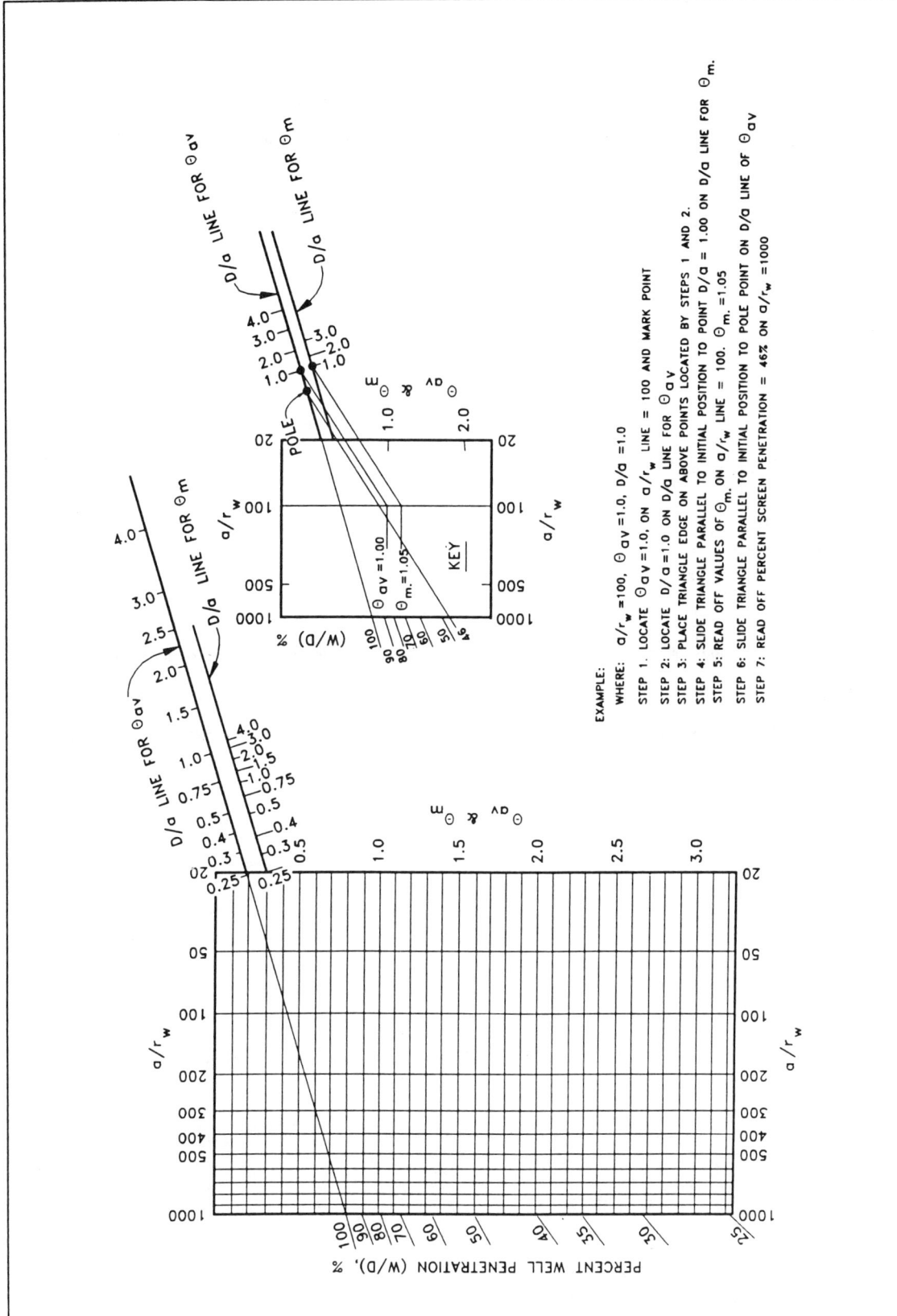

Figure 5-8. Nomographic chart for design of relief well systems

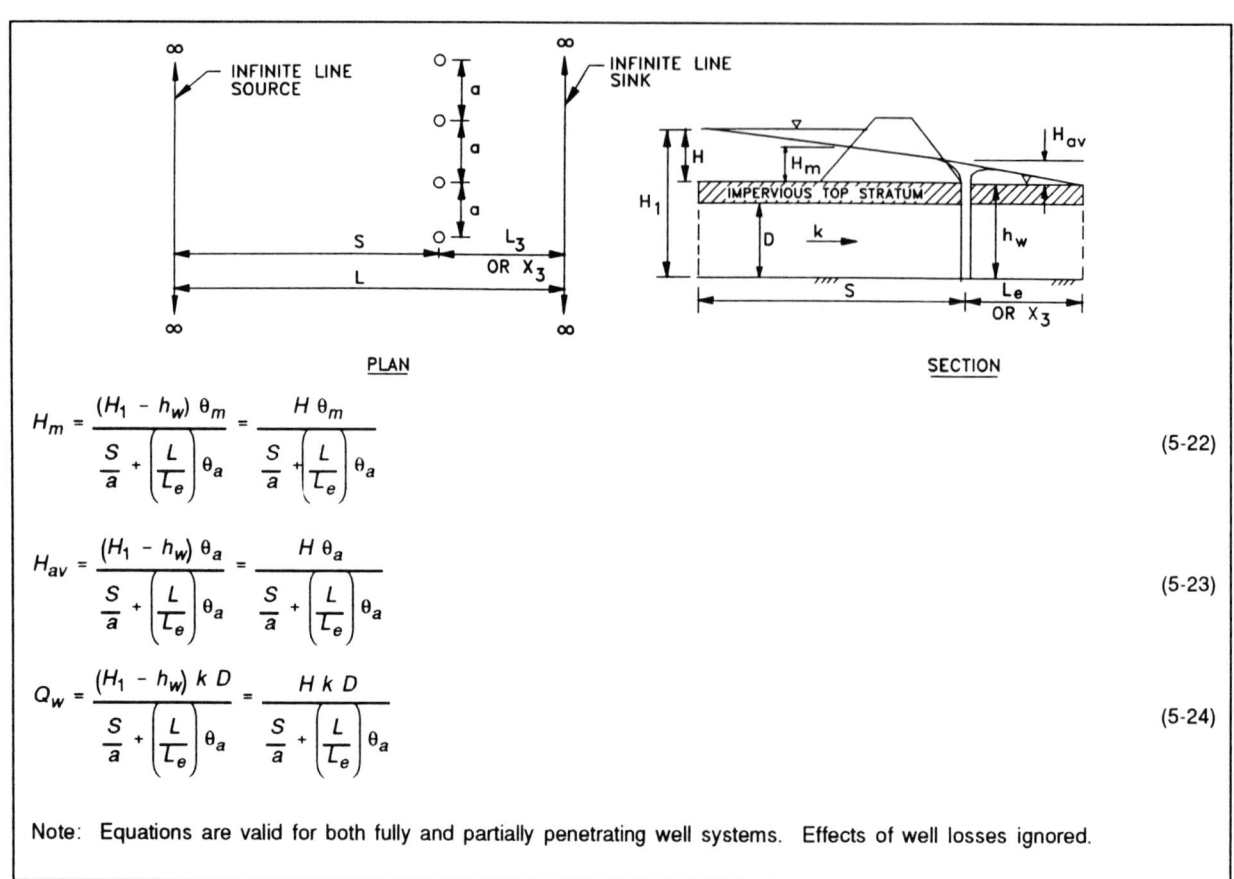

$$H_m = \frac{(H_1 - h_w)\,\theta_m}{\dfrac{S}{a} + \left(\dfrac{L}{L_e}\right)\theta_a} = \frac{H\,\theta_m}{\dfrac{S}{a} + \left(\dfrac{L}{L_e}\right)\theta_a} \qquad (5\text{-}22)$$

$$H_{av} = \frac{(H_1 - h_w)\,\theta_a}{\dfrac{S}{a} + \left(\dfrac{L}{L_e}\right)\theta_a} = \frac{H\,\theta_a}{\dfrac{S}{a} + \left(\dfrac{L}{L_e}\right)\theta_a} \qquad (5\text{-}23)$$

$$Q_w = \frac{(H_1 - h_w)\,k\,D}{\dfrac{S}{a} + \left(\dfrac{L}{L_e}\right)\theta_a} = \frac{H\,k\,D}{\dfrac{S}{a} + \left(\dfrac{L}{L_e}\right)\theta_a} \qquad (5\text{-}24)$$

Note: Equations are valid for both fully and partially penetrating well systems. Effects of well losses ignored.

Figure 5-9. Infinite line of wells parallel to infinite line source, impervious top stratum of finite length

ANALYSIS OF MULTIPLE WELL SYSTEMS

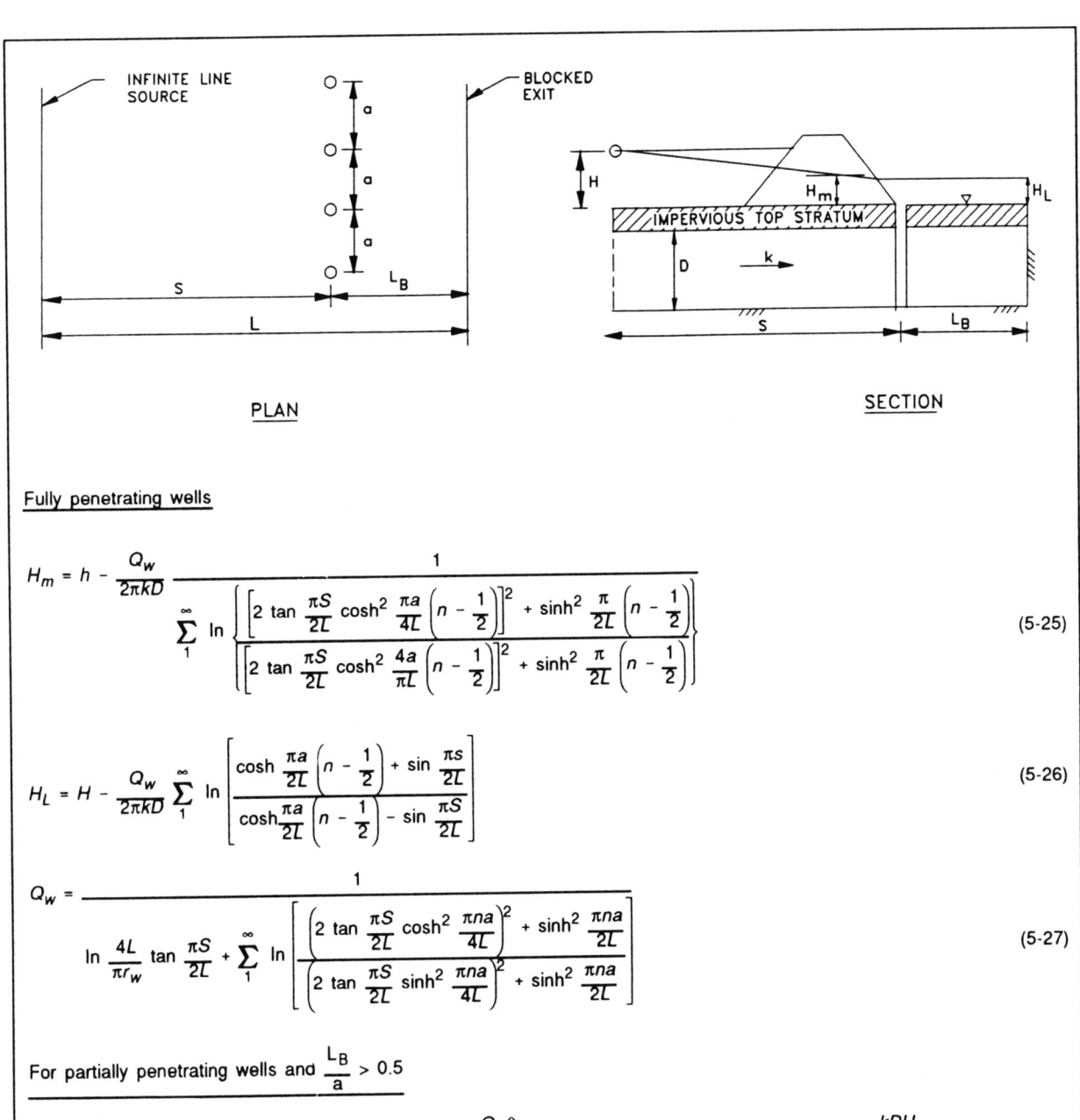

Figure 5-10. Infinite line of wells with blocked landside exit

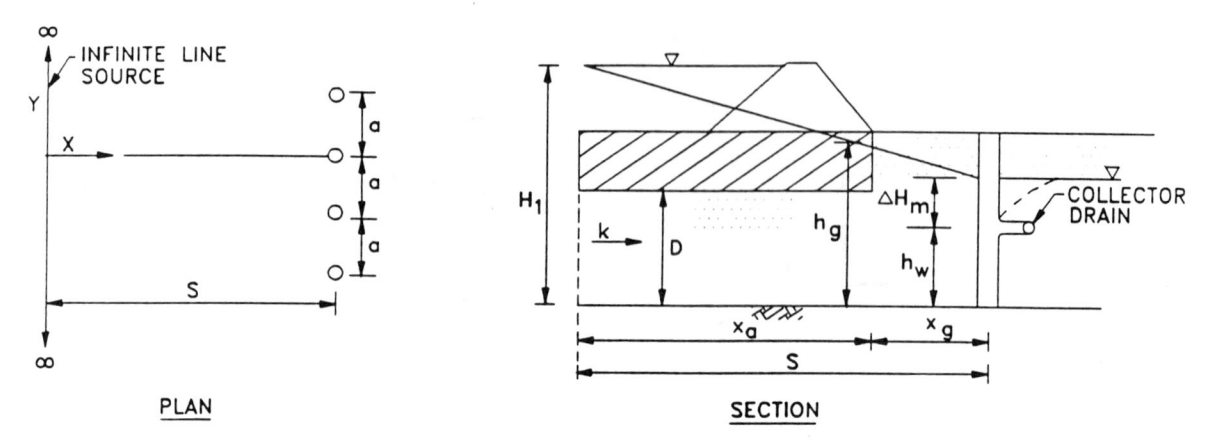

Gravity flow, Q_g = artesian flow Q_a

$$Q_g = \pi k \frac{\left(h_g^2 - h_w^2\right)}{\ln \frac{ae^{\frac{2\pi x_g}{a}}}{2\pi r_w}} = Q_a = \frac{k(H_1 - h_g)aD}{x_a} \qquad (5\text{-}31)$$

$$h_g = -\frac{\beta}{2} + \sqrt{\frac{\beta^2}{4} + \beta H_1 + h_w^2} \qquad (5\text{-}32)$$

where $\beta = \dfrac{aD}{\pi x_a} \ln\left(\dfrac{ae^{\frac{2\pi x_g}{a}}}{2\pi r_w}\right)$ (5-33)

h_p = height of saturation at any point in gravity flow zone

$$h_p = \sqrt{h_g^2 - \frac{\left(h_g^2 - h_w^2\right)}{\ln \frac{ae^{\frac{2\pi x_a}{a}}}{2\pi r_w}} \ln\left[\frac{\cosh\frac{2\pi}{a}(x + x_a) - \cos\frac{2\pi y}{a}}{\cosh\frac{2\pi}{a}(x - x_a) - \cos\frac{2\pi y}{a}}\right]} \qquad (5\text{-}34)$$

ΔH_m = excess head above the well outlet midway between wells

$$\Delta H_m = \sqrt{h_g^2 - \frac{\left(h_g^2 - h_w^2\right)}{\ln \frac{ae^{\frac{2\pi x_a}{a}}}{2\pi r_w}} \ln \frac{e^{\frac{2\pi x_a}{a}}}{2}} - h_w \qquad (5\text{-}35)$$

Figure 5-11. Infinite line of fully penetrating wells, combined gravity, and artesian flow

CHAPTER 6

WELL DESIGN

6-1. Description of Well

While the specific materials used in the construction vary and the dimensions and methods of installations differ, relief wells are basically very similar. They consist of a drilled hole to facilitate the installation; a screen or slotted pipe section to allow entrance of ground water; a bottom plate; a filter to prevent entrance and ultimate loss of foundation material; a riser to conduct the water to the ground surface; a check valve to allow escape of water and prevent backflooding and entrance of foreign material; backfill to prevent recharge of the formation by surface water; and a cover and some type of barricade protection to prevent vandalism and damage to the top of the well by maintenance crews, livestock, etc. Figure 2-1 shows a typical relief well installation. The hole is drilled large enough to provide a minimum thickness of 4 to 6 in. depending on the gradation of the filter material as subsequently described. The hole is also overdrilled in depth to provide for the fact that initial placements of filter material may be segregated. The amount of overdrilling required is variable depending upon the size of tremie pipe used for filter placement, the total depth of the well, and most importantly on the tendency of the selected filter material to segregate. The backfill indicated as sand in Figure 2-1 normally consist of concrete sand or otherwise excess filter material. Its only function is to fill the annular space around the riser pipe to prevent collapse of the boring; these granular materials are easily placed and require a minimum of compaction. The backfill indicated as concrete in Figure 2-1 forms a seal to prevent inflow of surface water from rains and flooding.

6-2. Materials for Wells

Commercially available well screens and riser pipes are fabricated from a variety of materials such as black iron, galvanized iron, stainless steel, brass, bronze, fiberglass, polyvinyl chloride (PVC), and other materials. How well a material performs with time depends upon its strength, resistance to damage by servicing operations, and resistance to attack by the chemical constituents of the ground water. Wood has proven to be very stable in most environments in well installations, as long as it is continuously submerged in water; however wood well screens and risers are no longer commercially available. Stainless steel is apparently a very stable material in most environments; however it is relatively expensive. Type 304 stainless steel has excellent corrosion resistance; whereas Type 403 stainless steel has moderate corrosion resistance. Low-carbon or other-type steel wire-wrapped screen may be more economical in many instances, however it has no corrosion resistance. Brass and bronze are extremely expensive and are not completely stable in some acid environments. Fiberglass is a promising material, however its performance history is relatively short. PVC appears to be completely stable, and it is easy to handle and install; however it is a relatively weak material and easily damaged. The life of iron screens is extended by galvanizing, which may not provide permanent protection. Ferrous and nonferrous metals should never be placed in direct contact with each other, such as the case of a brass screen and a steel riser. The direct contact of these dissimilar metals may induce electrolysis and a resultant deterioration of the material.

6-3. Selection of Materials

Since pressure-relief wells are designed and installed to protect the foundations of structures, selection of materials for the well should be based on costs and performance over the life of the structure which it protects. Generally, the choice of well screen material will depend on three factors: (a) water quality, (b) potential presence of iron bacteria, and (c) strength requirements. A water quality analysis will determine the chemical nature of the ground water and indicate whether it is corrosive and/or incrusting (see Table 3-1). Enlargement of screen openings due to corrosion can cause progressive movement of fines into the well, therefore it is essential that the well screen be fabricated from corrosion-resistant material where corrosive waters are expected. Similarly, if incrusting ground water is expected, future maintenance which may require acid treatments as described in Chapter 12 necessitates the use of material that can withstand the corrosive effect of the treatments. When the presence of iron bacteria is anticipated, the well screen should be selected which can withstand the damaging effects of the repeated chemical treatments described in Chapter 12. The strength of the well screen is usually not a major factor when commercial well screens designed for deeper well installations are employed. The screen sections should be able to withstand maximum compression and tensile forces during installation operations as well as horizontal forces which may develop during installation and possibly later because of lateral earth movements.

6-4. Well Screen

A. SLOT TYPE. A variety of slot types are available in most types of well screens. PVC screens with open slots of varying dimensions consisting of a series of saw cuts are typically available. Metal and fiberglass screens are available with open slots, louvered or otherwise shielded slots, or "continuous slots." The "continuous slot" screens consist of a skeleton of vertical rods wrapped with a continuous spiral of wire. The wire can be a variety of cross-sectional shapes. The trapezoidal-shape wire provides a slot that is progressively larger toward the inside of the screen. This shape allows any filter gravel that enters the slot to fall into the well rather than clog the screen. The open-type slots are advantageous in developing the filter. They allow the successful use of water jets, whereas shielded slots deflect the water jet and reduce or destroy its effectiveness in the filter. Machine cut slots typically have jagged edges which facilitate the attachment of iron bacteria making screens difficult to treat later. Continuous slot screens are commercially fabricated of Type 304 and 316 stainless steel, monel, galvanized or ungalvanized low-carbon steel, and thermoplastic materials, mainly PVC and ABS or alloys of these materials. Couplings and the bottom plate for the well screen may be either glued, threaded, or welded and should be constructed of the same material as the well screen.

B. DIMENSIONS. The size of the individual openings in a well screen is dictated by the grain size of the filter. The openings should be as wide as possible, yet sufficiently small to minimize entrance of filter materials. Criteria for selection of screen opening size are presented subsequently. The anticipated maximum flow of the well dictates both the minimum total open-slot area of the screen (the spacing and length of slots) and the minimum diameter of the well. The open area of a well screen should be sufficiently large to maintain a low entrance velocity of less than 0.1 ft per second (fps) at the design flow. Representative areas and maximum well capacities for various well diameters with different continuous slot sizes are shown in Table 6-1. Well screen manufacturers should be consulted for more specific information. The well diameter must be large enough to conduct the maximum anticipated flow to the ground surface and facilitate testing and servicing of the well after installation. Head loss in the well should also be taken into consideration in selecting a well diameter.

6-5. Filter

A. In order to prevent infiltration of foundation sands into the filter, the filter gradation must meet the stability requirement that the 15 percent size of the filter should be not greater than five times the 85 percent size of the foundation materials. As shown in Figure 6-1, the design should be based on the finest gradation of the foundation materials, excluding zones of unusually fine materials where blank screen sections should be provided. If the foundation consists of strata with different grain size bands, different filter gradations should be designed for each band. Each filter gradation must also meet the permeability criterion that the 15 percent size of the filter should be more than three to five times the 15 percent size of foundation sands. Either well-graded or uniform filter materials may be used. A uniform filter material has a coefficient or uniformity, C_u, of less than 2.5 where C_u is defined as

$$C_u = \frac{D_{60}}{D_{10}} \quad (6\text{-}1)$$

where

D_{60} = grain size at which 60 percent by weight is finer

D_{10} = grain size at which 10 percent by weight is finer

The C_u of well-graded filter materials should be greater than 2.5 and less than 6 to minimize segregation. The grain sizes should be reasonably well distributed over the specified range with no sizes missing. Well-graded filter materials used with proper well development procedures increase efficiency and permit the use of large screen openings; however they are subject to

WELL DESIGN

Table 6-1. Properties of Wire-wrapped Continuous Slot Screens
(Manufactured by Johnson Division, SES inc.)

Shipping Weight		Intake Areas (square inches per foot of screen)							
		Slot Opening Size							
Lb/Ft	Nom. Diam.	10-slot	20-slot	40-slot	60-slot	80-slot	100-slot	150-slot	250-slot
4	3	15	26	41	52	59	65	73	82
5	3 1/2	18	31	49	61	70	77	88	99
6	4	20	35	57	71	81	88	101	115
6	4 1/2	23	40	64	80	92	100	114	129
7	5	26	45	72	90	102	112	112	132
8	5 5/8	28	49	79	99	113	123	141	159
10	6	30	53	85	106	100	112	132	156
15	8	28	51	87	113	133	149	160	194
19	10	36	65	108	141	166	186	200	243
22	12	42	77	130	143	171	195	237	265
35	14	37	68	97	132	161	185	232	292
41	16	42	60	108	148	180	208	261	327
47	18	36	69	124	169	206	237	298	375
57	20	41	77	139	189	229	264	280	366
71	24	61	113	131	182	226	265	343	449
72	26	63	118	138	191	237	278	360	471
81	30	75	138	161	224	278	325	422	552
91	36	84	157	184	255	317	371	481	629

Notes:
1. Open areas may differ somewhat from these figures. Extra-strong construction, for example, reduces open areas in some cases because heaver material is used to increase screen strength.
2. The maximum transmitting capacity of the screen can be derived from these figures. To determine gpm per ft of screen, multiply the intake area in square inches by 0.31. It must be remembered that this is the maximum capacity of the screen under ideal conditions with an entrance velocity of 0.1 fps.

to segregation during handling and placement. Well-graded filters should have an annular thickness of 6 to 8 in. Uniformly graded filters permit a lesser annular thickness of filter (4 to 6 in.) and are not subject to segregation, thereby reducing the amount of overdrilling.

B. The filter should consist of natural material made up of hard, durable particles. It should contain no detrimental quantities of organic matter or soft, friable, thin, or elongated particles. Crushed carbonate aggregates should be avoided because they tend to break down with a loss in permeability. Furthermore, they will tend to dissolve if the wells require future acid treatment as part of future rehabilitation operations. It is often difficult to purchase material that meets the required gradation, and it may be necessary to have the material specially blended. The special blends are expensive and sometimes difficult to acquire, but essential to the installation of acceptable permanent relief wells.

6-6. Selection of Screen-Opening Size

In general, the slot width (or hole diameter) of the screen should be equal to or less than the 50 percent size of the finest gradation of filter. Application of this criterion is demonstrated in Figure 6-1. Use of the 50 per cent size criterion for the selection of screen slot size appears to provide reasonable assurance against inwash of filter materials during well development and surging and furthermore results in suitably large openings to minimize the effects of incrustations and blockages which may develop during the life of the well (Hadj-Hamou, Tavassoli, and Sherman (1990).

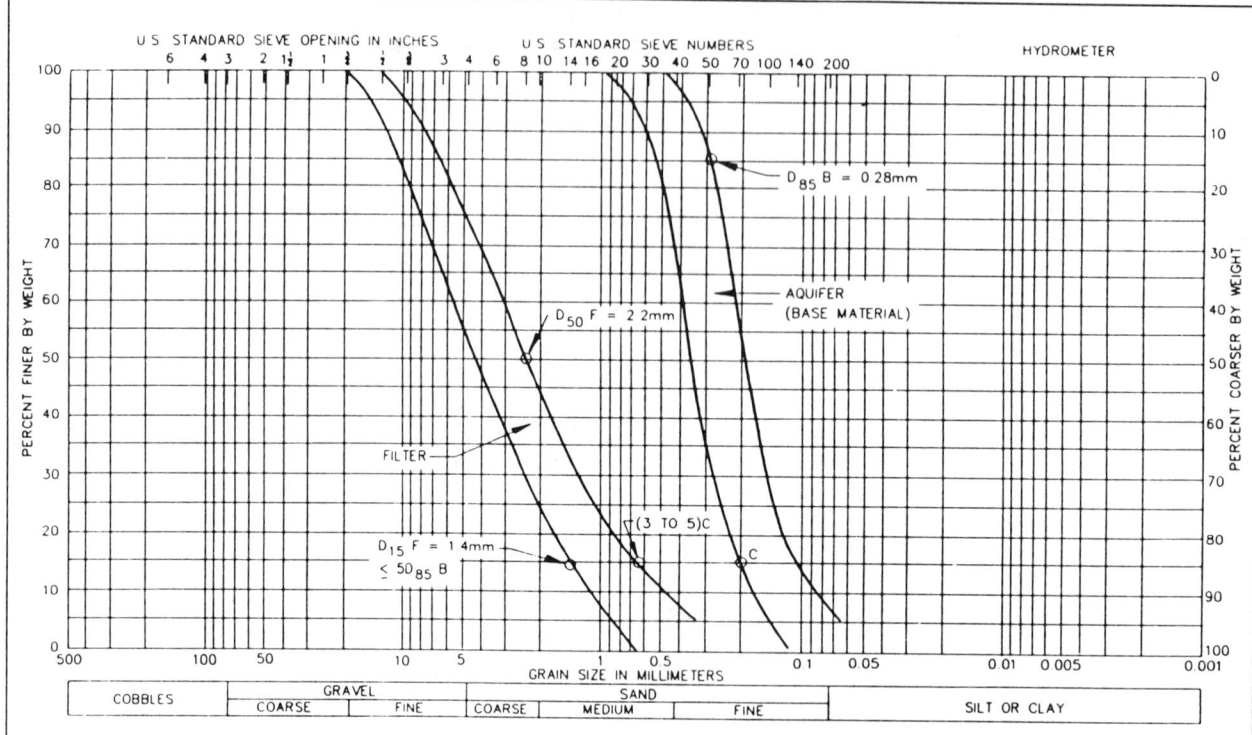

General Procedures
1. Determine minimum $D_{85}B$ on band of grain size curves for aquifer.
2. Determine maximum $D_{15}F$ for filter material based on the stability criterion. $D_{15}F < 5\, D_{85}B$
3. Select a widely graded ($C_u > 2.5$) or uniform filter material meeting the above criteria.
4. Establish a reasonable band of grain sizes for the filter.
5. Check to ensure that permeability criterion is satisfied, i.e. the D_{15} size of the filter band is 3 to 5 times greater than the D_{15} size of the aquifer band.
6. Select a maximum screen slot size equal to the D_{50} size of the fine curve of the filter band.

Figure 6-1. Typical design of filter for relief well

6-7. Well Losses

A. Head losses within the system consist of entrance head loss in the screen and filter (H_e) plus friction head losses arising from flow in the screen, riser, and connections (H_f) plus velocity head loss (H_v). The total hydraulic head loss in a well (H_w) is given by

$$H_w = H_e + H_f + H_v \qquad (6\text{-}2)$$

B. The entrance losses in the screen and filter for a properly designed and developed screen and filter will generally be relatively small at the time of well installation. Installation techniques resulting in smear or undue disturbance of the drill hole walls, however, can result in relatively large initial entrance losses. Entrance losses can be expected to increase with time for a variety of reasons discussed in Chapter 11. For example, as shown in Figure 6-2, the entrance losses for 8-in.-ID slotted wood well screens, based on piezometer data at the time of installation, amounted only to about 0.10 to 0.25 ft for a flow through the screen of 10 gpm per foot of screen. However, as shown in Figure 6-2, entrance losses for the particular wells increased significantly with time. The initial entrance losses for wire-wrapped screens should

WELL DESIGN

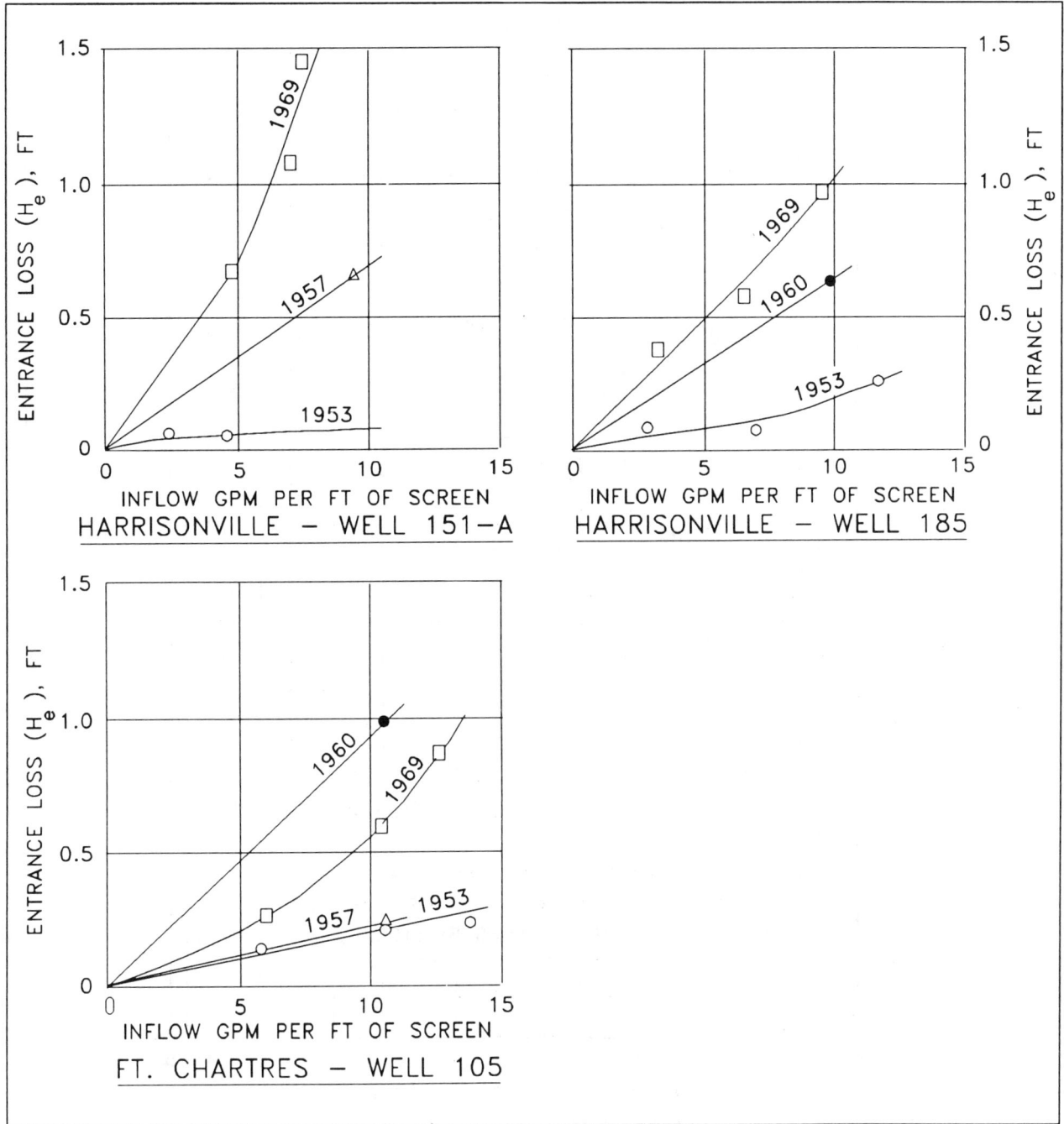

Figure 6-2. Entrance losses versus inflow for 8-in-ID slotted wood well screens in St. Louis District (after Montgomery 1972)

be even less. Both field and laboratory tests indicate that the average entrance velocity of water moving into the screen should not exceed 0.1 fps. At this velocity, friction losses in the screen openings will be negligible and the rates of incrustation and corrosion will be minimal. The average entrance velocity is calculated by dividing estimated well yield by the total area of the screen openings. If the velocity is greater than 0.1 fps, the screen length and/or diameter should be increased accordingly. The long-term value of entrance loss is difficult to predict, and unless experience in a specific location is available, conservative values based on Figure 6-2 should be selected.

c. Friction losses in the screen and riser sections may be estimated from Figure 6-3. The head loss in the screen section should be computed for a distance of one-half the screen length. More

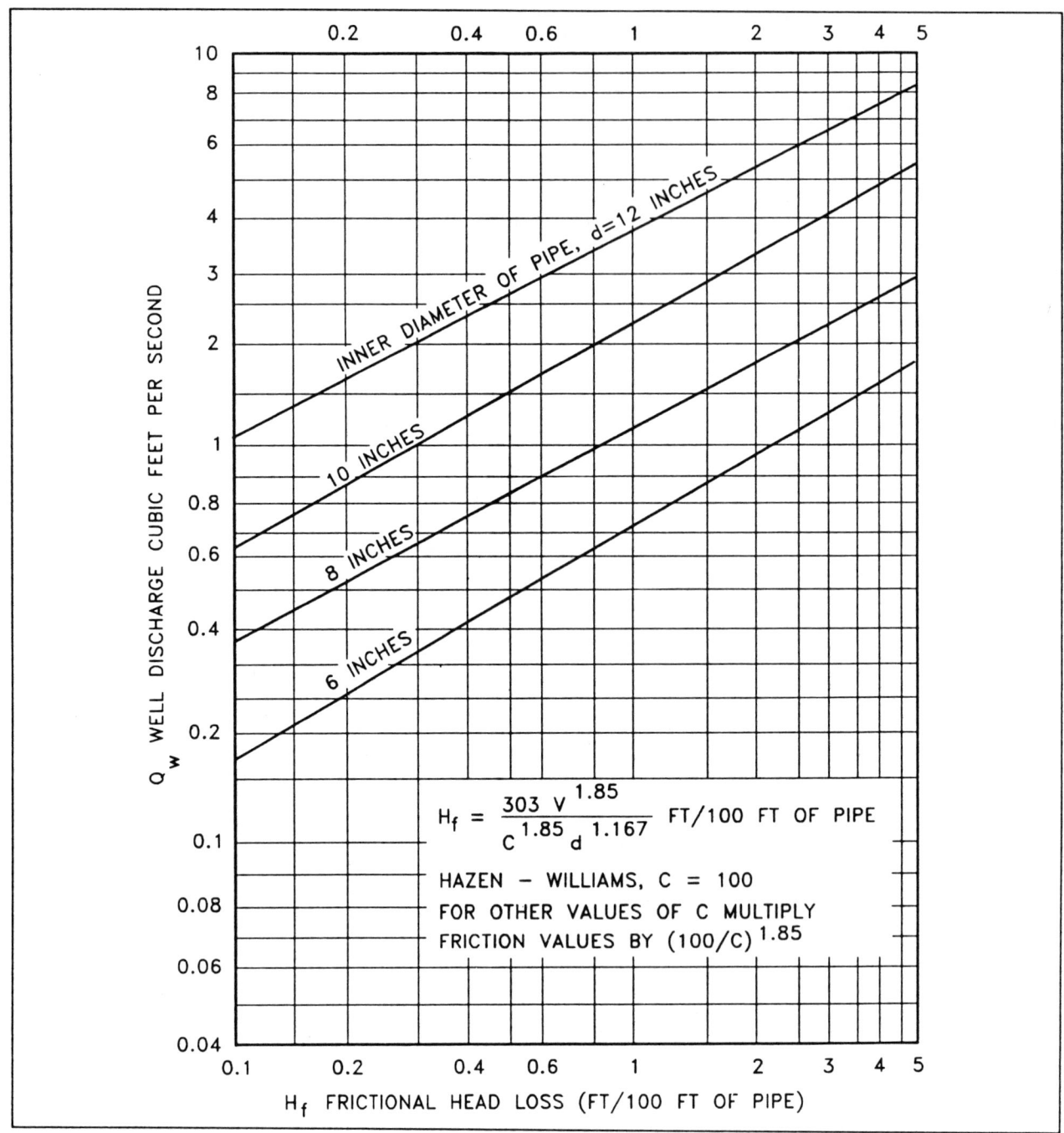

Figure 6-3. Friction-head losses in screen and riser sections

WELL DESIGN

accurately, friction losses can be calculated according to the Darcy-Weisbach formula as described in EM 1110-2-1602. The resistance coefficient in the formula is solved by the Colebrook-White equation also given in EM 1110-2-1602. This equation requires the input of an effective roughness parameter for the material comprising the well screen and riser pipe. A computer code for the solution of the Colebrook-White equation is given in USAEWES (1973).

D. Velocity head losses, H_v, should be computed by means of the equation

$$H_v = \frac{v^2}{2g} \qquad (6\text{-}3)$$

where

v = the velocity of the water in the riser pipe
g = acceleration due to gravity = 32.2 ft/sec^2

Losses due to elbow connections should be included where applicable.

6-8. Effective Well Radius

The effective well radius to be used in design computations is calculated as the outside radius of the well screen plus one-half the thickness of the filter.

CHAPTER 7

DESIGN OF WELL SYSTEMS

7-1. General Approach

The design of relief well systems consists essentially of determining the location and penetration of wells that will reduce the piezometric surface of the substratum pressure, h_o, in landside or downstream areas to an allowable head, h_a. Analyses are made using formulas presented in Chapters 4 and 5. Where wells are required along the toe of a levee or dam, the wells will generally be located along a line so that their locations are defined by a well-spacing. The well-spacing is first determined assuming an infinitely long line of wells, and then the spacing is reduced where necessary to allow for the reduced efficiency of a finite number of wells compared to the infinite number. For given boundary conditions and the same allowable head, there are any number of combinations of well-spacing and penetration that will suffice. The final selected spacing and penetration should be based to a great extent on the most economical design. The presence of natural topographic features may require adjustment in the design well-spacing to ensure that well outlets are located at the lowest practical elevation.

7-2. Design Heads

The design of relief well systems for dams are based on steady state conditions which would prevail with the reservoir pool at the maximum design level. This reservoir pool normally is taken as the top of the surcharge pool. The design net head is the difference between the latter elevation and downstream tailwater elevation, usually taken as downstream ground surface or lower, if appropriate. In the case of relief well design for levees, the design net head is usually taken as the difference in elevation between net grade of the levee and tailwater.

7-3. Boundary Conditions

Boundary conditions which must be determined include the distance to the effective source of seepage entry, S; the distance from the line of relief wells to the effective seepage exit, x_3; and the distance to a blocked exit, L_B, if such exists. Procedures for the determination of these values are given in Appendix B.

7-4. Design Procedures

Direct application of the formulas in Chapters 4 and 5 is not possible as they are based on the assumption that the hydraulic head losses in the well are zero. As shown in Figure 5-4, the head losses must be determined on the basis of the computed well flow and added to the maximum landside head with wells, which in turn would result in a lower factor of safety with respect to uplift. If the tops of the well risers extend above tailwater, the difference in elevation should be added to the well losses in determining the maximum landside head with wells. The maximum landside head will always occur midway between wells for fully penetrating wells. For partially penetrating wells, there may be a difference as the average head may exceed the head midway between wells. To maintain the required factor of safety, a reduction in well-spacing is required so as to lower the maximum landside head with wells. Thus, an iterative procedure must be utilized to find the well spacing which satisfies the condition that when well losses are considered, the head midway between wells or the average head, whichever is greater, equals the design values. Procedures involving this concept are presented below.

7-5. Infinite Line of Wells, Impervious Top Stratum

The general procedure for designing a system of relief wells along an infinite line with an impervious top stratum extending to a great distance landward follows. The procedure is valid for both fully and partially penetrating well systems.

A. Compute the allowable head, h_a, under the top stratum at the downstream toe of the dam or levee from Equation 3-2. Assume tailwater elevation coincides with ground surface (or as appropriate).

B. Assume that H_m, the net head midway between wells, is equal to h_a and that well losses, H_w, are equal to zero (see Figure 5-4a).

C. For a given well penetration, W, compute H_m for various trial values of well-spacing, a, based on Equation 5-14. Interpolate to determine the required well-spacing for $H_m = h_a$.

D. Calculate the well flow, Q_w, for the above well-spacing and penetration using Equation 5-17.

E. Assume the well dimensions, and calculate the well losses, H_w, corresponding to Q_w.

F. Repeat step **C** using $h_m = H_m - h_w$ in place of H_m in Equation 5-14, and determine a new value of a.

G. Repeat steps **D** through **F** until relatively consistent values of a are obtained on two successive trials. The value of a derived in this manner satisfies the design requirements for fully penetrating wells.

H. If the wells are not fully penetrating, repeat step **B** assuming that H_{av}, the average net head, is equal to h_a.

I. For a given well penetration, W, compute H_{av} for various trial values of well-spacing, a, based on Equation 5-15. Interpolate to determine required well-spacing for $H_{av} = h_a$.

J. Calculate the well flow, Q_w, for the above well-spacing and penetration using Equation 5-17.

K. Assume the well dimensions and calculate the well losses, H_w, corresponding to Q_w.

L. Repeat step **I** using $h_{av} = H_{av} - H_w$ in place of H_{av} in Equation 5-15 and determine a new value of a.

M. Repeat steps **J** through **L** until relatively consistent values of a are obtained on two successive trials. For design, select the lesser value of well-spacing determined from steps **G** and **L** for a given well penetration.

N. Repeat for various well penetrations to develop a relation between well penetration and spacing that satisfies design requirements.

7-6. Infinite Line of Wells, Wells in Ditch

The design of an infinite line of wells with the well outlets located in a collector ditch to lower landside ground-water levels below ground surface should be based on the assumption that the top stratum is impervious regardless of its permeability. The design procedure is essentially similar to that described in the preceding paragraph with the following exceptions:

A. Assume an elevation for the well outlets, h_w, and compute the allowable head, h_a, under the top stratum beneath the collector ditch from Equation 3-3. Proceed with steps **B** through **G** to obtain a design well-spacing that satisfies Equation 3-3.

B. Select a design ground-water level landward of the wells defined by $h_d = h_w + \Delta_D$. Assume $h_d = H_{av}$.

C. Proceed with step **I** using h_d in place of H_{av}.

D. Continue steps **J** through **N** to obtain the design well-spacing.

7-7. Infinite Line of Wells with Impervious Top Stratum of Finite Length

The procedure for design of an infinite line of wells with a landside impervious top stratum of finite length is presented below. The procedure is also applicable to the case of a semipervious landside top stratum after conversion to an equivalent length of impervious stratum as discussed in Appendix B. The spacing for an infinite line of relief wells for a given penetration is determined using an iterative procedure. For small well-spacings, the average uplift factor Θ_a will be equal to or larger than Θ_m and will control. For large well-spacings, Θ_m will be equal to or larger than Θ_a and will therefore control. A summary of the equations used is shown on Figure 7-1. The procedure for computing the well-spacing for both conditions is as follows:

A. Compute the allowable head beneath the top stratum at the downstream toe of the dam, h_a, from Equation 3-2.

B. Assume that the net head in the plane of the wells, H_{av}, is equal to h_a and calculate the net seepage gradient toward the well line, ΔM, substituting in Equation 7-6 as follows:

$$\Delta M = \frac{h - h_a}{S} - \frac{h_a}{x_3} \qquad (7\text{-}11)$$

where

S = distance from effective seepage entry to line of wells

x_3 = distance from line of wells at the landside toe to effective seepage exit (length of landside impervious top stratum).

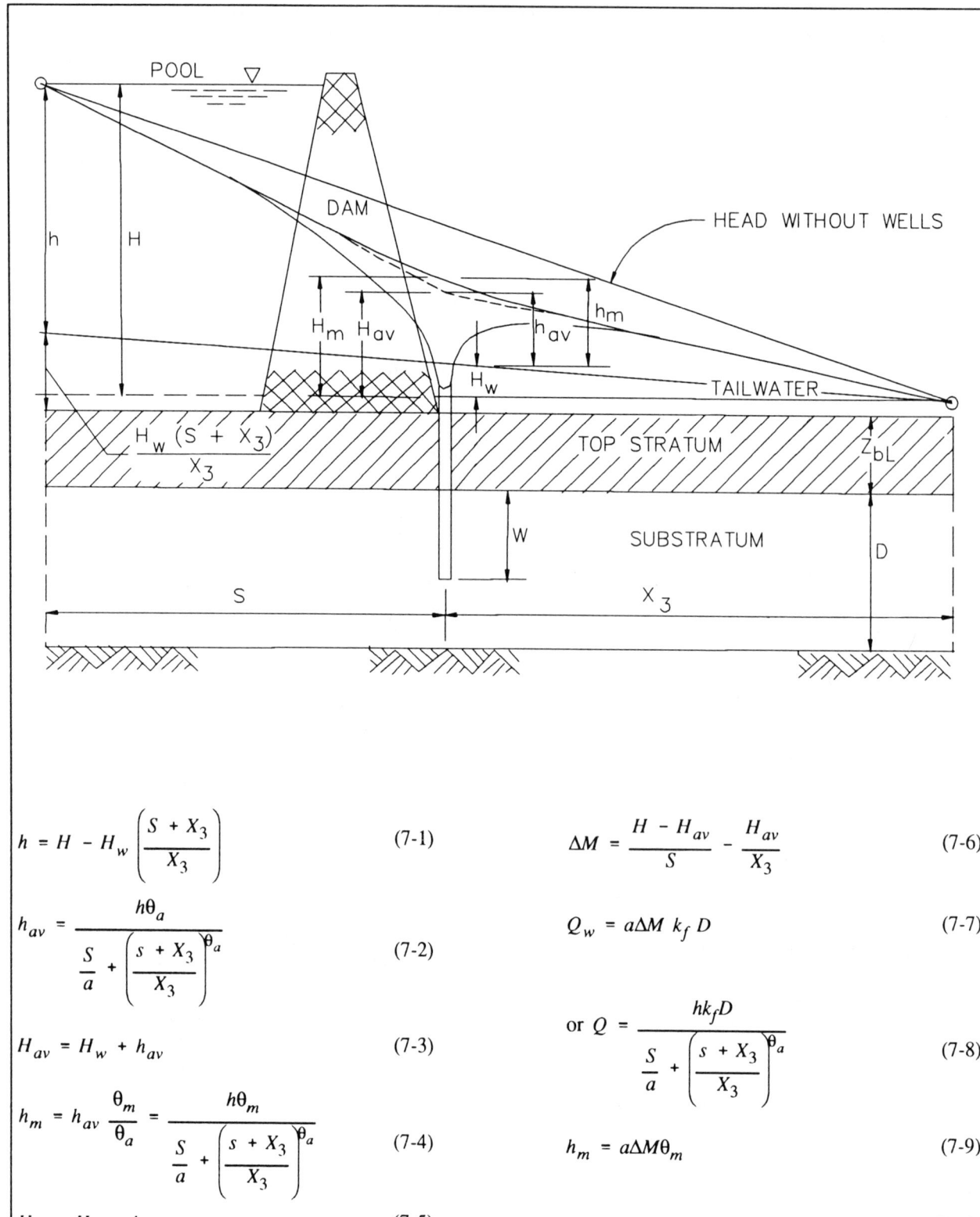

Figure 7-1. Nomenclature and formulas for design of relief wells at toe of dam

C. Assume a well-spacing and compute the flow from a single well using Equation 7-7.

D. Assume the well dimensions and calculate the well losses, H_w, corresponding to Q_w.

E. Compute the net average head in the plane of wells, h_{av}, using Equation 7-3.

F. Substitute values of ΔM and h_{av} in steps **B** and **E** and solve for Θ_a using Equation 7-10.

G. Find Θ_a from Table 5-1 and Figure 5-6 or Figure 5-8 for the given well penetration using the values of a used in step **F** and the corresponding a/r_w and D/a values.

H. The first trial well-spacing is that of value a for which Θ_a from step **F** equals Θ_a from step **G**.

I. Find Θ_m from Table 5-1 and Figure 5-7 or Figure 5-8 for the given well penetration and first trial well-spacing and the corresponding values of a/r_w and D/a values.

J. If $\Theta_a > \Theta_m$ repeat steps **C** to **I** using the first trial well-spacing in lieu of the spacing originally used in step **C**, and determine the second trial well-spacing. This procedure should be repeated until relatively consistent values of a are obtained on two successive trials. Usually the second trial spacing is sufficiently accurate. If in step **J**, $\Theta_a < \Theta_m$, a modified procedure is used for a second trial using steps **K** through **T**.

K. Assume $H_m = h_a$ and compute Q_w from Equation 7-7 using the value of ΔM obtained in step **B** and the first trial well-spacing from **H**.

L. Estimate H_w from Q_w of step **K**.

M. Compute the net head midway between the wells as $h_m = H_m - H_w$.

N. Using Θ_a and Θ_m from steps **H** and **I**, respectively, compute H_{av} from Equation 7-3.

O. Using H_w and h_{av} from steps **L** and **N** respectively, compute H_{av} from Equation 7-3.

P. Compute ΔM from Equation 7-6 using H_{av} from step **O**.

Q. Using h_m and ΔM from steps **M** and **P** respectively, compute Θ_m for various values of a from Equation 7-9.

R. Find Θ_m from Table 5-1 and Figure 5-7 or Figure 5-8 for the values of a used in step **Q** and the corresponding a/r_w and D/a values.

S. The second trial well-spacing is that value of a which Θ_m from step **Q** equals Θ_m from step **R**.

T. Find Θ_a from Figure 5-6 for the second trial well-spacing and the corresponding values of a/r_w and D/a.

U. Determine the third trial well-spacing by repeating steps **K** to **T** using the second trial well-spacing in lieu of the spacing originally assumed in step **K**, and in step **N** using the values Θ_m and Θ_a from steps **S** and **T** respectively, instead of those from steps **H** and **I**. This procedure should be repeated until relatively consistent values of a are obtained on two successive trials. Normally, the third trial is sufficiently accurate.

V. Repeat steps **G** through **U** for various well penetrations to develop a relation between well penetration and spacings that satisfies design requirements.

7-8. Computer Programs

A computer program for design of relief wells systems based on the above procedures was developed by Conroy (1984). Comparisons of the computer and hand solutions are presented by Cunny, Agostinelli, and Taylor (1989).

7-9. Head Distribution for Finite Line of Relief Wells

In a short, finite line of relief wells, the heads midway between wells exceed those for an infinite line of wells both at the center and near the ends of the well system as shown in Figure 7-2. With an infinite line of wells, the heads midway between wells are constant along the entire length of the well line. Many well systems may be fairly short; thus, it will be necessary to reduce the well-spacing computed for an infinite line of wells so that heads midway between wells will not be more than the allowable head under the top stratum. The ratio of the head midway between wells at the center of finite systems to the head between wells in an infinite line of wells, for various well-spacings and exit lengths, is given in Figure 7-3. The spacing of relief wells in a finite line should be the same as that required in an infinite line of wells to reduce the head midway between wells to h_a divided by the ratio of H_{m_n}/H_{m_∞} from Figure 7-3. In any finite line of wells of constant penetration and spacing, the head midway between wells near the ends of the system exceeds that at the center of a system. Thus, at the end of both short and long well systems, the relief wells should generally be made deeper to provide additional penetration of the previous substratum in order to obtain the same head reduction as in the central part of the well line. In the case of fully penetrating wells, the same head reduction can be obtained by additional wells using gradually decreasing well-spacings near the ends of the line. The above-mentioned procedures for designing

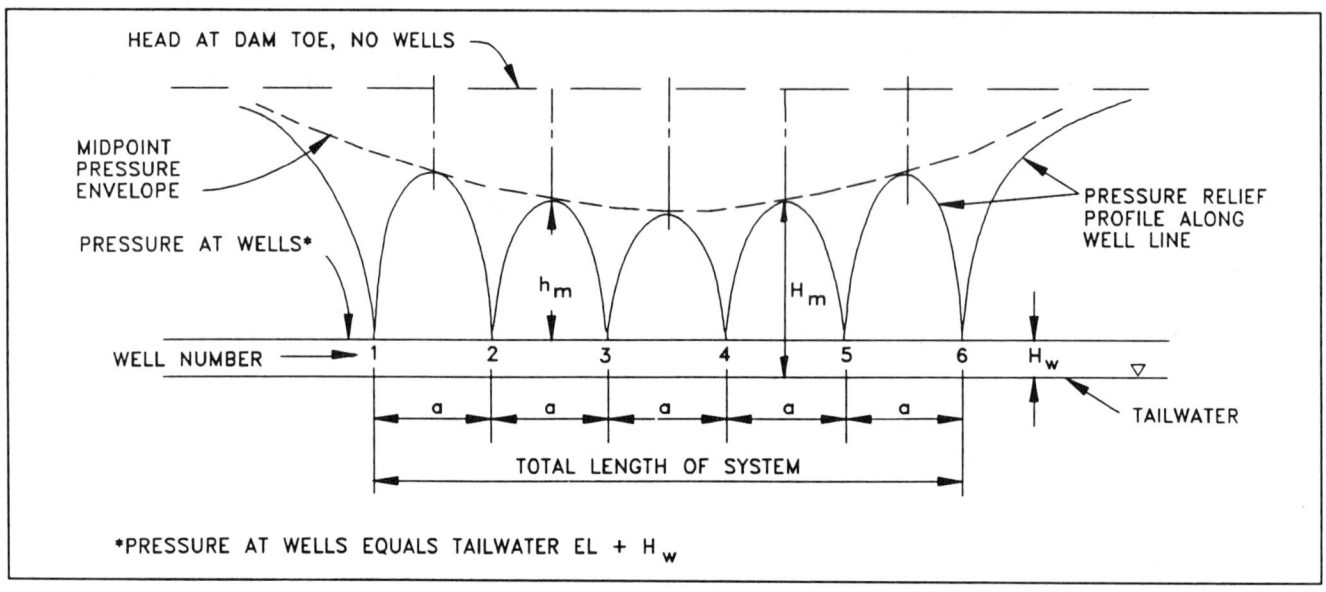

Figure 7-2. Variation of pressure relief along a finite line of relief wells (after EM 1110-2-1905)

finite relief well systems, although approximate, are usually sufficient.

7-10. Well Systems at Outlet Works and Spillways

When well systems for outlet structures and spillway structures are being designed, the problem is to design a group with a finite number of wells with proper spacing and penetration which will reduce the head at the center of the well group to the allowable design head. In this type problem, usually the pressure at the well is known because such wells normally will be discharging into tailwater elevation, possibly through a collector system either under the stilling basin or along the channel-side slopes. Thus, the head at the well is equal to tailwater plus the hydraulic head loss in the well and collector system. This elevation when subtracted from the reservoir pool represents the net head acting on the system. For such a system of fully penetrating wells, the equations using the method of images for fully penetrating wells with a line source and impervious downstream top stratum are utilized, and the head at the center of the well group is calculated by superposition.

7-11. Well Costs

The design of relief well systems will normally produce various combinations of well-spacing and penetration which satisfy the design criteria. The optimum design should be based on initial cost as well as overall costs including maintenance and possible replacement costs over the life of the structure. Costs should be calculated per 100-ft stationing as shown in Figure 7-4. Elements included in the estimate of initial costs are the cost of drilling or other installation technique, as well as the cost of well screen, riser pipe, and filter, all of which are on a foot basis. Additional fixed costs include back-filling, well development and testing, plus the costs of well guards, check valves, and horizontal outlet pipes if used. As shown in Figure 7-4, the well-spacing and screen penetration should be selected that will result in the minimum well cost per station over the life of the structure.

7-12. Seepage Calculations

As previously noted, the presence of relief wells will tend to increase the total quantity of seepage beneath a levee or dam, Q_s. The seepage per foot of structure with no wells is computed by the equation.

$$Q_s = \frac{kDH}{S + x_3} \qquad (7\text{-}12)$$

DESIGN OF WELL SYSTEMS

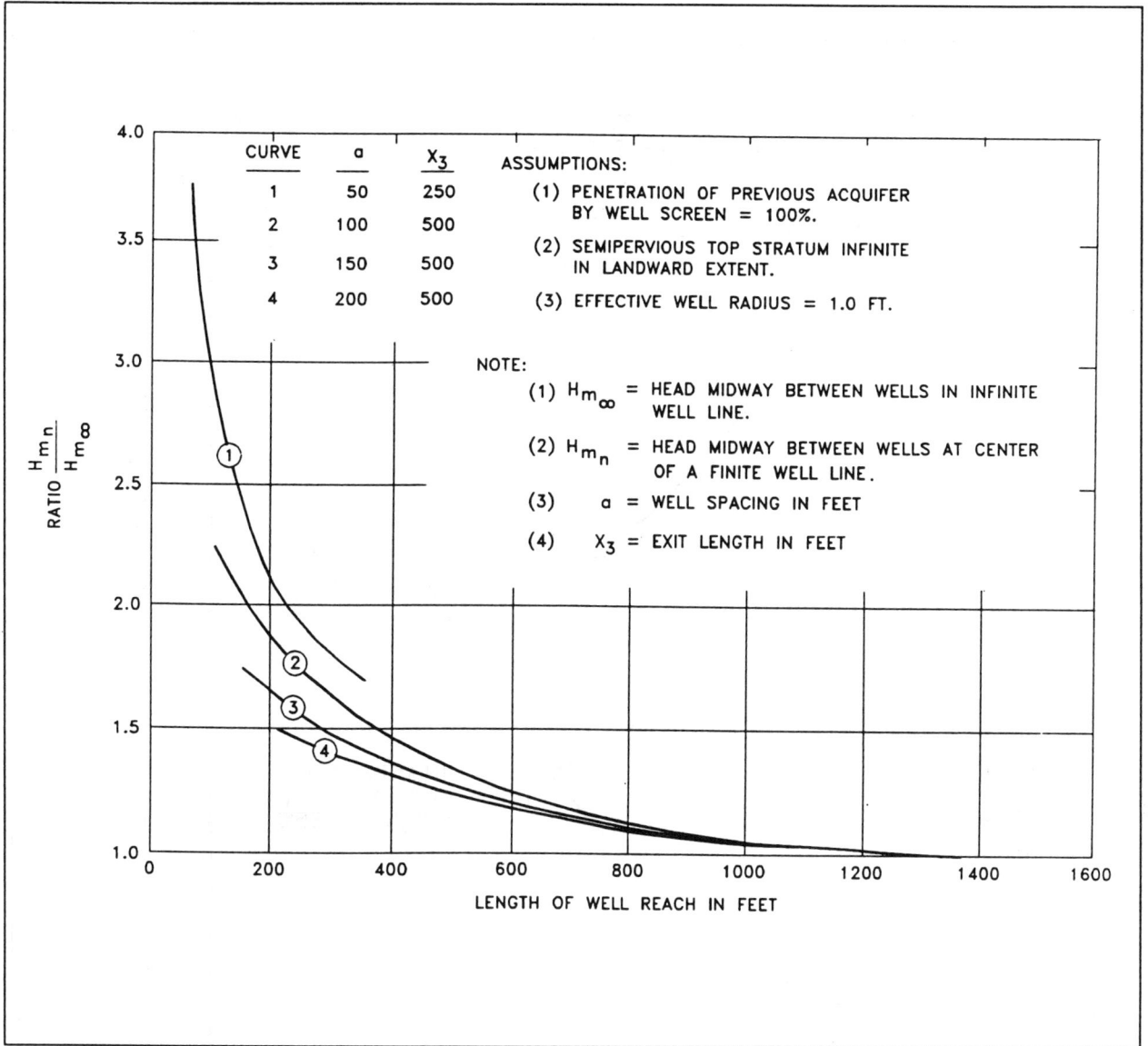

Figure 7-3. Ratio of head midway between relief wells at center of a finite well system to head midway between wells in an infinite system (after EM 1110-2-1905)

The seepage with wells is equal to the flow from the well system $\sum Q_w$, plus the seepage beyond the well system, Q_{sw}, which is computed by the equation

$$Q_{sw} = \frac{kDH_{av}}{x_3} \qquad (7\text{-}13)$$

where H_{av} is computed from equations in Figure 7-1. The estimation of the total quantity of seepage passing beneath a structure with or without wells can be of importance in selecting a well-spacing which will intercept the desired amount of seepage.

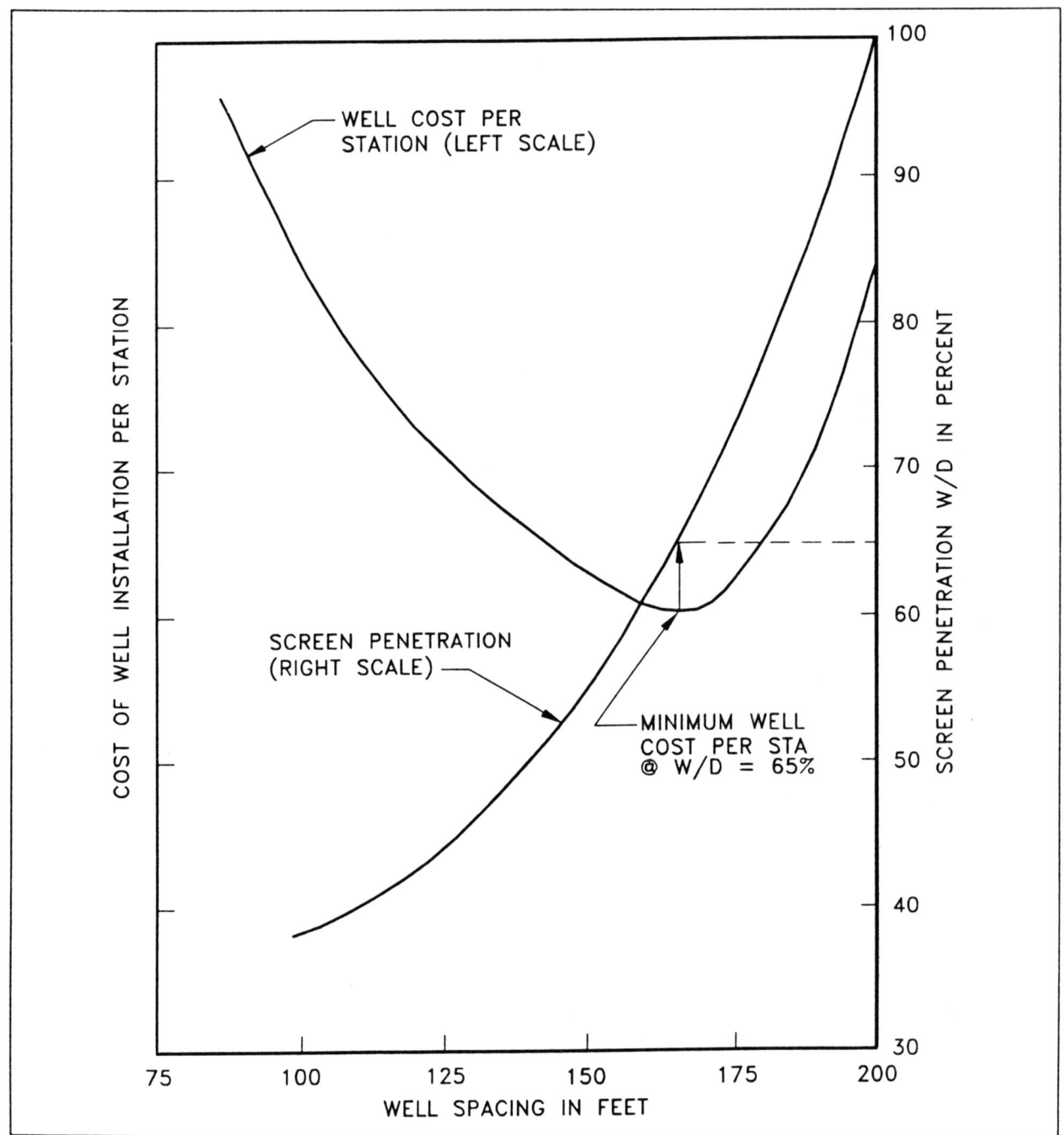

Figure 7-4. Determination of optimum well design

CHAPTER 8

RELIEF WELL INSTALLATION

8-1. General Requirements

Proper installation of relief wells is essential to the successful protection of the structures for which they are designed to protect. Before installation is begun, all materials required for completion of the installation should be on hand at the worksite. The well screen and riser should be checked for proper material, length, diameter, and slot openings. The filter material should be inspected and checked against gradation specifications. Successful completion of a well installation is often dependent upon time, and many installations have been aborted because of delays. An open boring of sufficient size and depth is necessary to facilitate the installation of a well. The hole should be vertical so that the screen and riser may be installed straight and plumb. As previously discussed, the hole is drilled large enough to provide a minimum thickness of 4 to 6 in., depending on the gradation, of the filter material. The methods of providing an open boring in the ground are numerous; however not all are acceptable for the installation of permanent relief wells, and those considered acceptable are discussed in the following paragraphs.

8-2. Standard Rotary Method

One method of drilling for well installation which has gained popularity in the well drilling industry is standard rotary drilling using a biodegradable, organic drilling fluid additive. No bentonitic clays are used in the drilling fluid. Standard rotary drilling consists of rotating a cutter bit against the bottom of a boring, while a fluid is pumped down through the drill pipe to cool and lubricate the bit and return the cuttings up the open hole to the ground surface. The required size of bit is governed by the screen diameter and the thickness of filter. The ability of the fluid to carry the cuttings is dependent on its velocity and viscosity. The velocity of the returning fluid is reduced with increased boring diameter, and the reduction is compensated by increased viscosity of the drilling fluid. One such drilling fluid additive is marketed under the trade name "Revert," so-called because the fluid reverts to the viscosity of water, normally in about three days. Chemicals can be added to speed up or delay the reversion of such fluids as "Revert." Ground-water temperatures may effect reversion times.

A. EQUIPMENT. A rotary-type drill rig of sufficient hoisting and torque capacity is required. The cutter or drill bit can be of either drag or roller design. The drill pipe should be as large as practicable to increase the volume of the fluid at the drill bit and, consequently, the velocity of the fluid returning up the open hole.

B. PROBLEMS. The reverting process of the drilling fluid leaves a small amount of slimy ash which, unavoidably, is mixed into the filter material, however a large percentage of this ash is removed during development of the well. Testing to determine the extent of detriment caused by this ash residue has not been sufficient to evaluate the effectiveness of this method, however it has been used successfully in installation of permanent relief wells. Chemical development of the well is required as subsequently described.

8-3. Reverse-Rotary Method

This method is generally considered to provide the best acceptable drill hole and should be used whenever possible for the installation of permanent relief wells. In the reverse-rotary method, the hole for the well is made by rotary drilling, using a similar cutting process as employed in standard rotary drilling except the drilling fluid is pulled up through the drill pipe by vacuum and the drilling fluid reenters the top of the open boring by gravity. Soil from the drilling is removed from the hole by the flow of drilling fluid circulating from the ground surface down the hole and back up the hollow drill stem from the bit. Since the cross-sectional area of the boring is many times larger than that of the drill pipe, the slow downward velocity of the fluid acting against the open boring does not erode the walls. The drilling fluid consists of water and, unavoidably, a small amount of the finer fraction of the natural material being drilled. A high velocity is attained with the fluid returning up through the drill pipe, thus eliminating the need for a high viscosity. The drill water is circulated by a centrifugal or jet-eductor pump that pumps the flow from the drill stem into a sump pit. As the hole is advanced, the soil particles settle out in the sump pit, and the muddy water flows back into the drill hole through a ditch cut from the sump to the hole. The sides of the drill hole are stabilized by seepage forces acting against a thin film of fine-grained soil that forms on the wall of the hole. A sufficient seepage force to stabilize the hole is produced by maintaining the water level in the hole at least 7 ft above the natural water table. Figure 8-1 shows schematically the circulating system for reverse-rotary drilling. No bentonitic drilling mud should be used because of gelling in the filter and aquifer adjacent to the well. If the hole is drilled in clean sands, some silt soil may need to be added to the drilling water to attain the desired degree of muddiness (approximately 3,000 ppm). A biodegradable organic drilling fluid additive such as "Revert" or equivalent may also be added to the drilling water to reduce water loss.

A. EQUIPMENT. Reverse-circulation rotary drilling requires somewhat specialized equipment, most of which is commercially available or easily fabricated. Any rotary-type drill rig large enough to handle the load and having sufficient torque capability can be adapted to circulate water through an eductor to create a vacuum on the drill pipe. Drill pipe and hoses should be of a constant inside diameter throughout the system to assure that material entering the system can be circulated completely through it. In alluvial deposits, a drag-type bit similar to the cutter head for a dredge is sufficient. Roller-type bits are commercially available for use in consolidated deposits. The eductor consists of a pipe Y with a nozzle fitted into one end of the Y.

B. PROBLEMS. It is necessary to maintain an excess hydrostatic pressure on the drill hole to stabilize the walls. In most materials, a minimum excess head of 7 ft is required and greater is desirable. When the static water level is very near

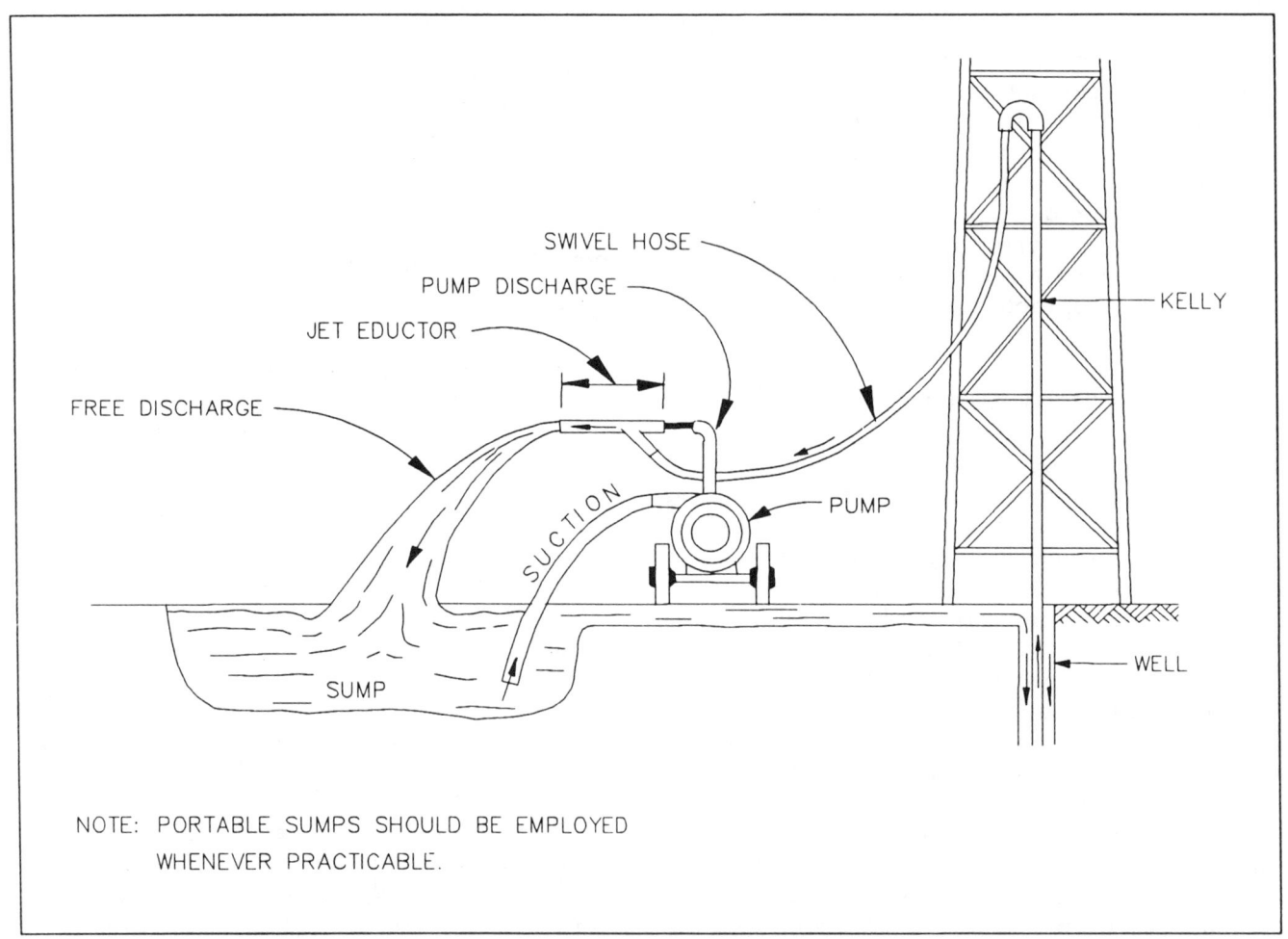

Figure 8-1. Schematic diagram of circulatory system (after EM 1110-2-1913)

the ground surface or artesian conditions prevail, it may be necessary to elevate the drilling rig on temporary beams. Some success has been experienced by lowering the water level with well points, but if the pressure is derived from a deeper, artesian source, it is necessary to lower the pressure in the aquifer with deep wells. Since the formation in which a well is installed consists predominately of granular material, the loss of water into the formation presents a problem during drilling. An almost unlimited supply of water can be necessary to maintain a completely filled, open boring. A large sump is required to supply adequate water. During the drilling, all cuttings from the boring are deposited in the sump and must be provided for. A sump three times the anticipated volume of the completed boring is adequate, if it can be kept filled with water from another source. Consideration should be given to the required thickness of the natural impervious clay blanket when constructing a sump. An instantaneous loss of water resulting in loss of excess head can cause failure of the boring walls. Often, if the rotation of the drill bit is stopped, the water loss is greatly reduced. The boring must be kept full of water until the well screen, riser, and filter are installed.

8-4. Bailing and Casing

In cases where standard or reverse-rotary drilling is not successful, an equally acceptable method of drilling consists of bailing while driving a steel casing into the hole to stabilize the boring walls. This method is economical in some materials, and it does not inject deleterious materials into the formation. Loose to medium dense, clean, granular materials can be bailed economically. Often the granular materials are overlain with a cohesive overburden which does not yield easily to bailing, and it is more economical to auger through this overburden.

A. EQUIPMENT. A drill rig with a wire line hoist and driving capability is adaptable to this method of well installation. It should be remembered that large casing, heavy enough to sustain driving, presents a sizable load to be handled by the drill rig. The use of a vibratory pile driver can greatly facilitate the driving and subsequent removal of the casing. The casing should be flush-joint, or welded-joint steel pipe. Two types of bailers are commonly used for this purpose (Figure 8-2). Details are given in EM 1110-2-1907. The bailer is operated on a wire line by lowering to the bottom of the boring and quickly pulling, or snatching, up a short distance a number of times to fill the bailer.

B. PROBLEMS. This method of drilling produces good results but often presents problems in operations. Thin layers of cohesive materials, or cemented materials within the formation, can preclude the advance by bailing and may also produce smear along the sides of the drill hole which could impair free flow into the well. Penetration of the casing can be retarded by friction of the granular formation against the outside of the casing unless vibratory hammers are used. After the casing is set, the boring completed, and the well installed, the casing is removed. The casing should be pulled, as the filter material is placed, to prevent disturbing the well installation by the friction of the filter material inside the casing. Using a vibratory pile hammer to drive and extract casing can densify loose foundation materials and filter materials. Generally, when material is densified, the hydraulic conductivity is reduced. The vibratory hammer cannot be used in wells that have more than one filter pack. As densification in the filter pack occurs, the material settles. This settlement, combined with settlement which occurs as the filter fills the void left by removal of the casing, results in uncertainties regarding the final position of the top of the filter. There are many uncertainties associated with this method of installation which makes it very difficult to estimate time and costs.

8-5. Bucket Augers

Under certain conditions drill holes for relief wells can be made with a bucket auger. The method has been successfully employed where cobbles up to 10 in. have been encountered. A bucket with side cutters is employed, and only water is used as the drilling fluid. The rate at which the bucket is inserted or withdrawn must be carefully controlled; thus close inspection is obligatory. A steel casing is installed through the top stratum to prevent smearing of fine-grained materials on the walls of the drill hole.

8-6. Disinfection

Before drilling begins, all tools, rods, bits, and pumps should be thoroughly washed with a chlorine solution to kill any bacteria remaining from previous well installations. Water used in the drilling process and filter materials should also be treated with a chlorine solution (Driscoll 1986). The strength of the chlorine solution should not be less than 100 ppm, which means a proportion of 100 lb of chlorine to 1 million lb of water. Calcium hypochlorite which contains 65 percent available

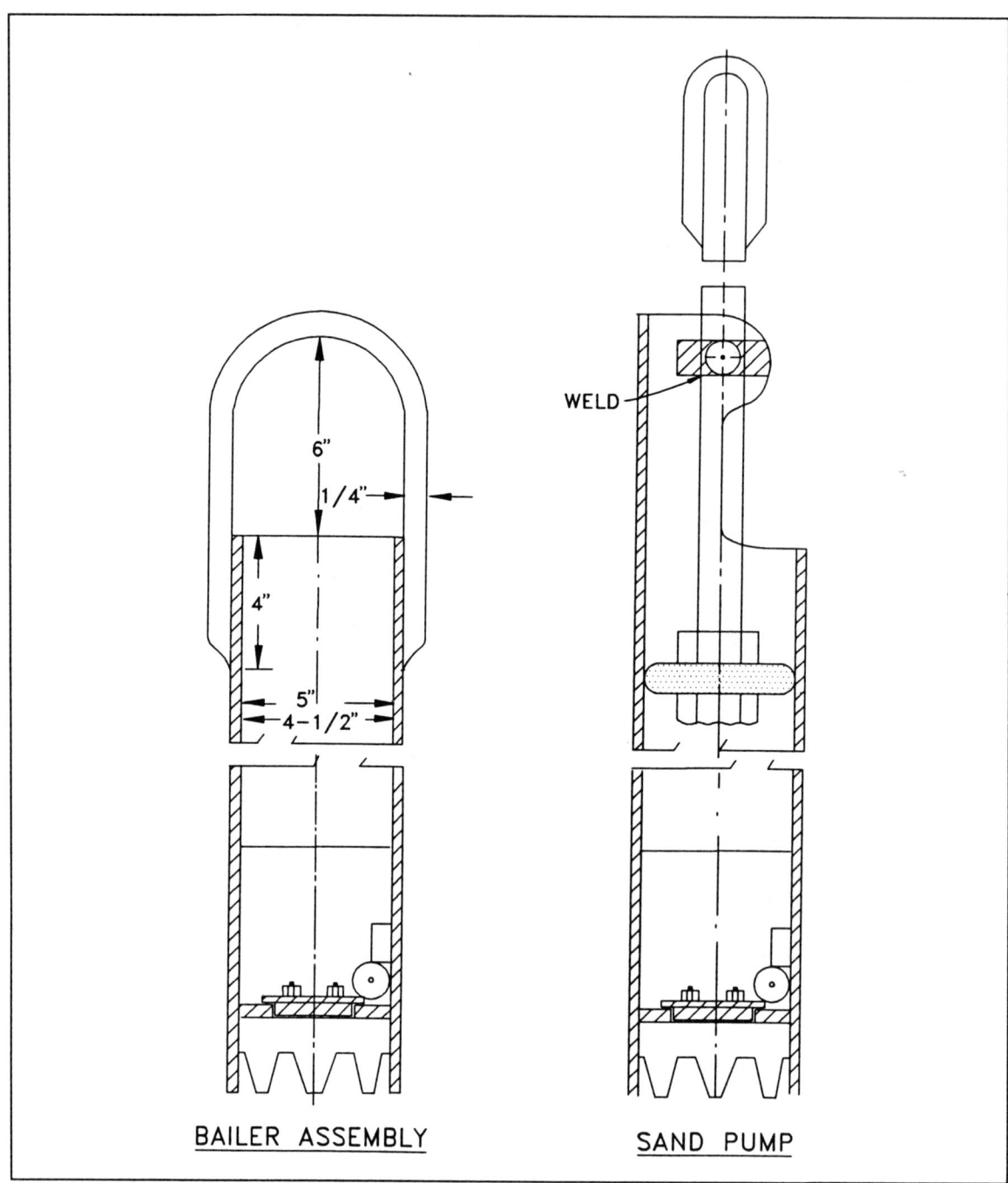

Figure 8-2. Bailer and sand pump assemblies (after EM 1110-2-1907)

chlorine is commonly used for this purpose. The required weight (wt) of calcium hypochlorite to produce a given strength in N gallons (gal) of water is given by the equation

$$Wt\,(lb) = N\,(gal) \times 8.33 \times \frac{solution\ strength}{available\ chlorine} \quad (8\text{-}1)$$

where both solution strength and available chlorine are expressed as a decimal. Thus, for a chlorine solution of 100 ppm in 1,000 gal of water, using calcium hypochlorite with 65 percent chlorine, the required weight of calcium hypochlorite is

$$Wt\,(lb) = 1,000\,(gal) \times 8.33 \times \frac{0.0001}{0.65} = 1.28\ lb$$

Similarly, for chlorine products such as sodium hypochlorite which is available in gallons, the required volume (V) to produce a given strength in N gal of water is given by the equation

$$V\,(gal) = N\,(gal) \times \frac{solution\ strength}{available\ chlorine} \quad (8\text{-}2)$$

Thus, for a chlorine solution of 100 ppm in 1,000 gal of water using sodium hypochlorite with 10 percent available chlorine, the required volume of sodium hypochlorite is

$$V\,(gal) = 1,000\,(gal) \times \frac{0.0001}{0.10} = 1.0\ gal$$

8-7. Installation of Well Screen and Riser Pipes

Once the boring is completed and the tools withdrawn, the boring should be sounded to assure an open hole to the proper depth. The well screen and riser pipe can be fabricated at the factory in varying lengths. The contractor will determine these lengths based on the capacity of his equipment. The bottom joint of the well screen should be fitted with a cap or plug to seal the bottom of the screen. The lengths of screen are connected together as they are lowered into the hole. Each length must be measured to determine its total made-up length, and the bottom of the screen should be set at the designed depth, or as field conditions require. The method of connecting the lengths of screen and riser vary: metal screen and riser have threaded or welded joints; plastic and fiberglass screens usually have either mechanical or glued joints. Each joint should be made up securely to prevent separation of the well during installation and servicing activities. Each joint should be kept as straight as possible to facilitate ease of servicing and testing. The riser and screen sections of the well should be centered in the drill hole by means of appropriate centering devices to facilitate a continuous filter around the well screen. If materials appreciably finer than anticipated in design are encountered, design personnel should be notified. In such cases, it may be necessary to replace the screen by a solid pipe or blank screen to prevent piping of foundation materials into the well. Immediately after installation of the well screen and riser, the total inside depth should be sounded. The exact inside depth of the well must be known to determine whether damage occurs during development and servicing of the well.

8-8. Filter Placement

Caution in proper design, control of manufacture, and handling of filter materials to the jobsite can be completely negated by improper placement in the well. Acceptable construction of permanent relief wells demands that the filter be placed without segregation because widely graded filters when placed in increments tend to segregate as they pass through water, with coarse particles falling faster than fine particles. A tremie should be used to maintain a continuous flow of material and thus minimize segregation during placement. A properly designed, uniform (D_{90}/D_{10}<3 to 4) filter sand may be placed without tremieing if it is poured in around the screen in a heavy continuous stream to minimize segregation. The tremie pipe should be at least 2 in. in diameter, be perforated with slots 1/16 to 3/32 in. wide and about 6 in. long, and have flush screw joints. The slots allow the filter material to become saturated, thereby breaking the surface tension and preventing "bulking" of the filter in the tremie. One or two slots per linear foot of tremie is generally sufficient. To avoid contamination by iron bacteria, the filter should be washed through the tremie pipe using a 100-ppm chlorine solution. The tremie pipe is lowered to the bottom of the open drill hole, outside the well screen and riser pipe. The presence of centering devices will interfere with the proper use of the tremie by preventing uniform filling to some extent. The use of dual diametrically opposed tremie pipes will ensure more uniform placement. After the tremie pipes have been lowered to the bottom of the hole, they should be filled with filter material and then slowly raised to keep them full of filter material at all times. Extending the filter material at least 2 ft above the top of the screen will depend on the depth of the well to compensate for settlement during well development. The top of the filter should also terminate below the bottom of the overlying top stratum if present. The level of drilling fluid or water in a reverse-rotary drilled hole must be maintained at least 7 ft above the natural ground-water level until all the filter material is placed. If a casing is used, it should be pulled as the filter material is placed, and the bottom of the casing kept 2 to 10 ft below the top of the filter material.

8-9. Development

A well is at best inefficient until properly developed. Development procedures include both chemical and mechanical processes. Development of a well should be accomplished as soon after the hole has been drilled as practicable. Delay in doing this procedure may prevent a well being developed to the efficiency assumed in design.

8-10. Chemical Development

Chemical development is applied usually in the case where special drilling fluids are utilized and chemicals are injected into the well to aid in the dissolution of the residual drilling fluid in the filter. The chemicals should be of a type and concentration recommended by the manufacturer of the drilling fluid. They should be placed starting at the bottom of the well and dispersed throughout the entire screen length by slowly raising and lowering the injection pipe. After the chemicals have been dispersed, the well should be pumped and the effluent checked to ensure that the drilling fluid has completely broken down.

8-11. Mechanical Development

The purpose of mechanical development is to remove any film or silt from the walls of the drilled hole and to develop the filter immediately adjacent to the screen to permit an easy flow of water into the well. The result of proper development is the grading of the filter from coarsest to finest extending from the well. The effect of proper development is an increase in the effective size of the well, a reduction of entrance losses into the well, and an increase in the efficiency of the well. Many factors, including but not limited to development methods, well design, and filter installation, affect the time it takes to fully develop a well. Basically there are three methods used in development as discussed below.

A. WATER JETTING. A water jet, consisting of a series of small nozzles at the end of a pipe, lowered into the well screen, is very effective in developing the continuous slot-type, wire-wrapped screens. A typical water jet is shown in Figure 8-3. Water is pumped down and out through the nozzles at a high velocity. Nozzles are directed toward the screen slots in small concentrated areas, as shown in Figure 8-4. The water jet equipment can be fabricated in local welding shops. The size and number of nozzles must be consistent with the size and length of the pipe through which the water is pumped to ensure a high-pressure and high-velocity jetting action. This method requires a high-pressure, relatively high-volume water pump. The lowest effective nozzle velocity for water jetting is about 100 fps. Better results are obtained with nozzle velocities between 150 and 300 fps. Normally, development with a water jet is started at the bottom of the screen. Jetting is accomplished at one depth with the jet rotated for a fixed period of time. The jet is raised approximately 0.5 ft; rotation and jetting is continued for another fixed period of time. For the most effective jetting, the wells should be pumped or airlifted during jetting to remove the fines as they are dislodged by the jetting. This process is continued until the entire well screen has been jetted. The jetting tool should be continuously in motion since a small amount of sand is disturbed and may cause localized erosion of the screen. Jetting must be repeated a number of times to ensure optimum development of the well.

B. SURGING. A surging block is a plunger consisting of one or more stiff rubber or leather discs attached to a heavy shaft. These discs should be about 1 in. smaller in diameter than the screen ID. A typical surge block is shown in Figure 8-5. Surging consists of moving water in and out of the screen using the up and down motion of the surge block through short sections of the well screen. The well should always be pumped or bailed to ensure a relatively free inflow of water prior to surging. Surging should begin with a slow and gentle motion above the well screen and continue with more vigor from the top of screen downward. This method is less effective than the water jet described above in continuous slot screens and more effective in screens with widely separated slots and louvered or shielded

RELIEF WELL INSTALLATION

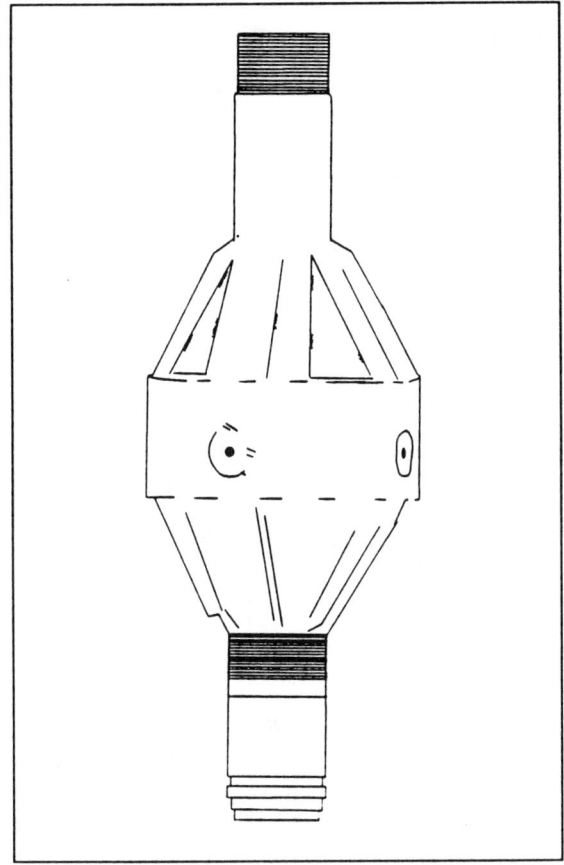

Figure 8-3. Schematic of four-nozzle jetting tool designed for use inside 8-in. well screen for jet development

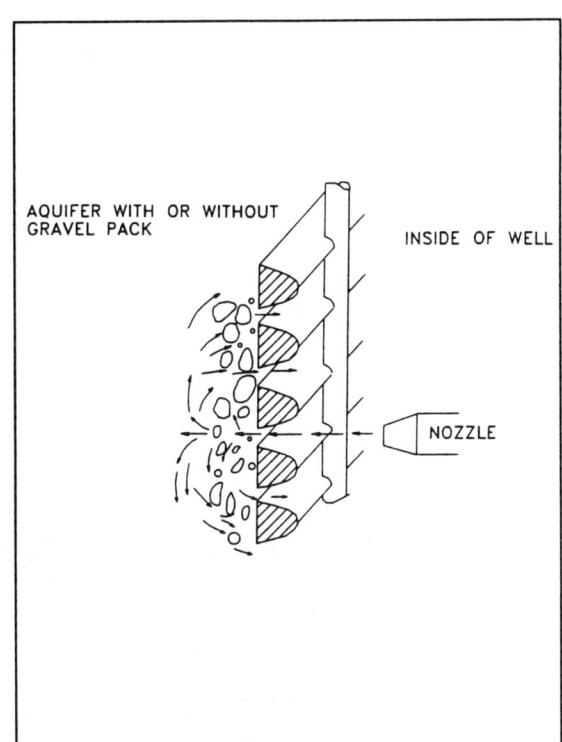

Figure 8-4. Well development by high-velocity jetting. Reprinted with permission of Wheelabrator Engineered Systems, Inc.

slots. The surging block should be pulled at approximately 2 fps for effective surging. For record keeping purposes, it is convenient to use 15 round trips as one cycle. The amount of material deposited in the bottom of the well should be determined after each cycle (about 15 trips per cycle). Surging should continue until the accumulation of material pulled through the well screen in any one cycle becomes less than about 0.2 ft deep. The well screen should be bailed clean if the accumulation of material in the bottom of the screen becomes more than 1 to 2 ft at any time during surging, then recleaned after surging is completed. Material bailed from a well should be inspected to see if any foundation sand is being removed. If the well is oversurged, the filter may be breached with resulting infiltration of foundation sand when the well is pumped.

C. PUMPING. One of the least effective and slowest methods of developing a well is simply pumping from the well. Pumping should be accomplished at a sufficient rate to effect maximum drawdown in the well.

The water passing from the formation

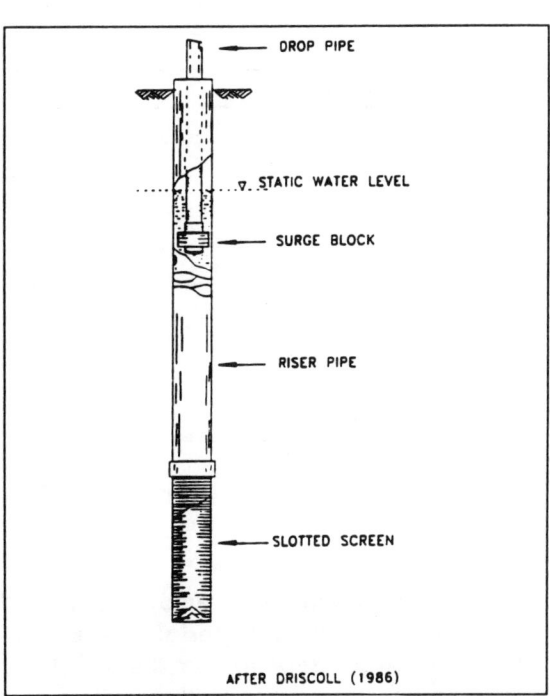

Figure 8-5. Development with a surge block. Reprinted with permission of Wheelabrator Engineered Systems, Inc.

RELIEF WELL INSTALLATION REPORT

PROJECT: | LEVEE DISTRICT:

LOCATION (STA): | OFFSET FROM CENTER LINE: | WELL NO.:

DRILLING METHOD: | DATE AND TIME:– FROM: TO:

WELL SCREEN:–TYPE: INSIDE DIAM: PERFORATIONS:

RISER:–TYPE: INSIDE DIAM: | FILTER:

TYPE OF WELL GUARD: INSIDE DIAM: | TYPE OF BACKFILLLL:

CONTRACTOR: | CONTRACT NO.: | FOREMAN: OPERATOR:

GROUND ELEV AT WELL: | DEPTH TO:– Fine Sand: Medium to Fine Sand:

ELEV TOP OF RISER: | ELEV TOP OF WELL GUARD: | ELEV TOP OF CONCRETE BACKFILL:

TYPE OF CHECK VALVE: | WELL EXTENSION:– Type: ;Ht:

FINAL WELL INSTALLATION DATA

RISER PIPE LENGTH:	INSIDE DEPTH OF WELL:
"EXTRA" SCREEN LENGTH:	DEPTH OF SAND IN WELL AFTER CLEANING:
"DESIGN" SCREEN LENGTH:	ELEV TOP OF WELL SCREEN:
TOTAL SCREEN LENGTH:	ELEV TOP OF FILTER:
"BLANK" PIPE LENGTH:	INSIDE BOTTOM ELEV OF WELL:
DEPTH OF HOLE:	ELEV RISER SET AT:

SURGING DATA

METHOD: | DATE AND TIME: | TOTAL SURGING TIME:

GRAVEL IN WELL BEFORE SURGING: | MATERIAL SURGED. IN:– | SURGING CYCLE #1: #2: #3:

SURGING CYCLE #4: #5: #6: #7: #8: TOTAL:

SETTLEMENT OF FILTER DURING SURGING:

PUMPING DATA

RATE OF PUMPING: | TOTAL PUMPING TIME:

RATE OF SAND INFILTRATION:–START: ;End:

REMARKS:

LOG OF HOLE	DEPTH	PIPE INSTALLED
	0	
	10	
	20	
	30	
	40	
	50	
	60	
	70	
	80	
	90	
	100	

WES FORM 797
MAR '53
REVISED OCT. '53

INSPECTOR

Figure 8-6. Relief well installation report 28

through the filter into the well removes part of the finer fraction of the filter material. The pumping equipment required depends on the size, yield, and anticipated drawdown in the well. Surging produced by repeatedly starting and stopping a pump is only effective where the static water level is well below the ground surface. Pumping, continued over a long period of time, is a reasonably effective method of well development. Pumping of the well is normally accomplished by inserting a pipe in the well and forcing compressed air to the bottom of the well. If the depth of submergence of the pipe is at least 50 percent of its length, air bubbles reduce the weight of the water column and will cause a flow to the ground surface. If 50 percent submergence is not possible, the water column which must be physically blown out of the well as it accumulates will require a large supply of air. Pumping can be accomplished using a mechanical pump, but granular material in the water can cause damage.

8-12. Sand Infiltration

During the development process, sand and silt will be brought into the well. When the depth of sand collected in the bottom of the screen reaches 1 ft, it should be removed by bailing. The accumulation of sand in the screen prevents development of that portion of the screen. A properly developed well will not produce an appreciable amount of sand, and entrance losses through the filter will be reduced to a minimum. In each of the methods discussed above, the actual amount of development must be recorded: the length, diameter, speed, and number of cycles of a surging block; the volume, pressure, and diameter of water jets; and the rate and method of pumping and length of time pumped. In addition, the amount of filter and foundation materials brought into the well and bailed out should be recorded. Upon completion of the development of the well, all material infiltrated into the well should be bailed out.

The well should be pumped to achieve a drawdown in the order of 5 ft in the well. If the well produces sand during pumping in excess at approximately 2 pints per hour (as determined from sounding and from collection of well flow in a 10-gal container) the well should be resurged or developed further and repumped. Wells continuing to produce excessive amounts of sand after 4 to 8 hours of surging or pumping should be abandoned and properly plugged.

8-13. Testing of Relief Wells

Performance of relief wells properly installed and developed is determined by pumping tests. The pumping test is used primarily to determine the specific capacity of the well and the amount of sand infiltration experienced during pumping. The information from this test is required to determine the acceptability of the well and will be used to evaluate its performance and loss of efficiency with time. The results of this pumping test must be made a part of the permanent record concerning the well.

A. EQUIPMENT. The equipment required for a pumping test consists of a pump of adequate size to effect a substantial drawdown. If the water level in the well is near enough to the ground surface, and the specific capacity of the well is high enough to produce a substantial flow with a small drawdown, a centrifugal pump may be used for this purpose. If the water level in the well is lower than about 18 to 20 ft, a deep-well pump will be required to effect substantial drawdown. A flow meter is required to measure the flow rate. A flat-bottom sounding device and a steel tape are required to determine the amount of sand infiltration deposited in the bottom of the well. A suitable baffled stilling basin is used to determine the amount of sand in the effluent. A sounding device suitable for determining the depth to the top of the water is needed to find the exact drawdown in the well. A well flow meter is desirable to measure the amount of flow at various depths within the well to define flow from various zones.

B. PUMPING. The well must be pumped to obtain a specified drawdown or flow rate. Drawdown measurements in the well should be made to the nearest 0.01 ft and recorded with the flow rate at 15-minute (min) intervals throughout the duration of the tests. Sufficient sand infiltration determinations are necessary to establish an infiltration rate for each hour of the pumping test. The rate of sand infiltration may be determined from sounding and measurements of sand in the effluent. For most properly developed wells, the amount of sand deposited in the well will be negligible and sand infiltration in the effluent can be recorded in terms of parts per million (Note: sand infiltration in parts per million is approximately equal to pints per hour times 3,000 divided by the pumping rate in gallons per minute) as measured with a centrifugal sand tester or other approved sediment concentration test (Driscoll 1986). The length of time that the pumping test must be continued is normally specified for the particular project. If the rate of sand infiltration during the last 15 min of the pumping test is more than 5 ppm, the well should be resurged by manipulation of the test pump for 15 min; then the test pumping should be resumed until the sand infiltration rate is reduced to less than 5 ppm. If after 6 hours (hr) of pumping the sand infiltration rate is more than 5 ppm, the well should be abandoned.

8-14. Backfilling of Well

After completion of the well testing, the annular space above the top of the filter gravel should be filled with filter gravel if necessary to achieve design grade. The remainder of the hole

RELIEF WELL PUMPING TEST REPORT

PROJECT:	LEVEE DISTRICT:	
LOCATION (STA):	ELEV TOP OF RISER:	WELL NO:
DATE:	TIME TEST STARTED:	TIME TEST COMPLETED:

TIME	ELAPSED TIME MINUTES	DEPTH TO WATER	DRAWDOWN IN FEET	FLOW IN GPM	TIME	ELAPSED TIME MINUTES	DEPTH TO WATER	DRAWDOWN IN FEET	FLOW IN GPM
	0		0						

SAND INFILTRATION TEST

DEPTH OF WELL:	DEPTH TO SAND IN WELL BEFORE TEST:	SAND IN WELL BEFORE TEST:

TEST NO.	TIME	DEPTH TO SAND (FT)	SAND IN WELL (PTS)	GAIN OR LOSS OF SAND IN WELL (PTS)	SAND PUMPED OUT OF WELL (PTS)	TOTAL INFLOW OF SAND INTO WELL (PTS)	LENGTH OF TEST (MIN)	RATE OF SAND INFILTRATION (PTS, HR)

DEPTH TO SAND IN WELL AFTER TEST:	SAND IN WELL AFTER TEST:	SAND IN WELL AFTER CLEANING:
REMARKS:		

WES FORM 796
MAR '53
REVISED OCT '53

INSPECTOR

Figure 8-7. Relief well pumping test report

should be filled with either a cement-bentonite mixture tremied into place or concrete where the height of drop does not exceed 8 ft. In both cases, a 12-in. layer of concrete sand or excess filter material should be placed on top of the filter before placement of grout or concrete. A tremie equipped with a side deflector will prevent jetting of a hole through the sand and into the filter.

8-15. Sterilization

Upon completion of the pumping tests and before installation of the well cover, each well should be sterilized by adding a chlorine solution with a minimum strength of 500 ppm. Sufficient solution should be added to the bottom of the well to provide a volume equal to three times the volume of the well based on the outer diameter of the filter. Before the solution is introduced into the well, all flow from the well should be stopped with inflatable packers or riser extensions. The solution should be injected into the well through a jetting tool by slowly raising and lowering the tool through the screened portion of the well. The well should be gently agitated at 10-min intervals every 2 hr for the first 8 hr and then at 8-hr intervals for at least 24 hr. As the chlorine will dilute with time, the concentration should be periodically checked. If it falls below 500 ppm, additional chlorine compound should be added. It should be noted that calcium hypochlorite may combine with naturally-occurring calcium in the ground water to form a precipitate of calcium hydroxide which can plug the pores of the foundation soils. Therefore, chlorine in the form of calcium hypochlorite should not be used in waters containing high calcium content.

8-16. Records

Permanent records of the installation, development, testing, and sterilization of a permanent relief well must be kept for evaluation of future testing. To monitor the efficiency and performance of the installation, the record must include identification of the well, method of drilling, type, length and size of well screen, and slot size. The filter should be defined as to grain-size characteristics, depth, and thickness. Elevation of the top of the well and the ground surface should be recorded. An abbreviated log of the boring should be included to define the depth to granular material, the thickness of that material, and the percent penetration of the well. Development data should include the method of development, the amount of effort expended in development, and the amount of materials pulled into the well during development. The record should show the final sounded depth of the well in case some fines remain at the bottom. The pumping test data should include the rate of pumping, the amount of drawdown, the length of time the pumping test was conducted, and the amount of sand infiltration during pumping. Installation and pumping test data should be recorded on forms similar to that shown in Figures 8-6 and 8-7. Forms should be filled in completely at the time each operation is completed and any additional observations should be recorded in a "remarks" section.

8-17. Abandoned Wells

Wells that produce excessive amounts of materials during pumping tests or that do not conform to specifications and can not be rehabilitated should be abandoned. Abandoned wells should be sealed to eliminate physical hazards, prevent contamination of ground water, conserve hydrostatic heads in aquifers, and prevent intermingling of desirable and undesirable waters. Primary sealing materials consist of cement or cement-bentonite grout placed from the bottom upward. In general, abandoned wells should be sealed following procedures established by local, state, or Federal regulatory agencies.

CHAPTER 9

RELIEF WELL OUTLETS

9-1. General Requirements

Relief wells should always be located where they are accessible by a drill rig for pump testing and cleaning and provided with outlets for this purpose. The outlets should be designed to minimize maintenance and to provide protection against contamination from backflooding, damage from floating debris, and vandalism. When wells are to discharge into a collector ditch or backwater which may contain organic matter, debris, and fine-grained sediment in suspension, or where high velocities may be expected while the wells are flowing, they should be installed off to the side and should discharge into the ditch or area through a tee connection and horizontal outlet pipe protected against corrosion. A flat-type check valve should be installed on the well riser with a flap gate on the end of the horizontal pipe. An example of this type of installation is shown in Figure 9-1.

9-2. Check Valves

Control of backflooding, which greatly impairs well efficiency, is best implemented by flat-type check valves constructed of aluminium (see Figure 9-1). The check valve is supported by a soft rubber gasket which fits snugly over the top of the riser or cast iron tenon set in the concrete backfill. Other types of check valves may be used but should be thoroughly tested under controlled conditions before application in the field.

9-3. Outlet Protection

For wells discharging at ground surface, the tops of the wells should be provided with a metal screen to safeguard against vandalism, accidental damage, and the entrance of debris. Details of a conventional metal well guard are shown in Figure 9-2. A suitable alternative consists of a section of stainless steel wire wound screen as shown in Figure 9-3. In the case of a T-type well where the top of the riser pipe is more than 5 ft below ground, the well guard should be 42 in. in diameter to permit safe access by a ladder. A guard screen consisting of a wire mesh with 1 in.-square openings may be installed at the end of the outlet pipe to prevent animals and debris from entering the outlet pipe in the event the flap gates do not close properly.

9-4. Plastic Sleeves

Where relief wells are provided for underseepage control at levees, the well flows at relatively low river stages will be somewhat in excess of natural seepage. In cases where the additional seepage is considered objectionable, each well can be provided with a plastic sleeve, 1.0 or 1.5 ft in length, which will raise the discharge elevation of the well accordingly. The sleeves prevent well flow at low river stages when no pressure relief is necessary. At higher river stages or as soon as substratum pressures develop to the extent that water begins to spill over the top of the sleeves, they should be removed so that the well can function as intended.

RELIEF WELL OUTLETS

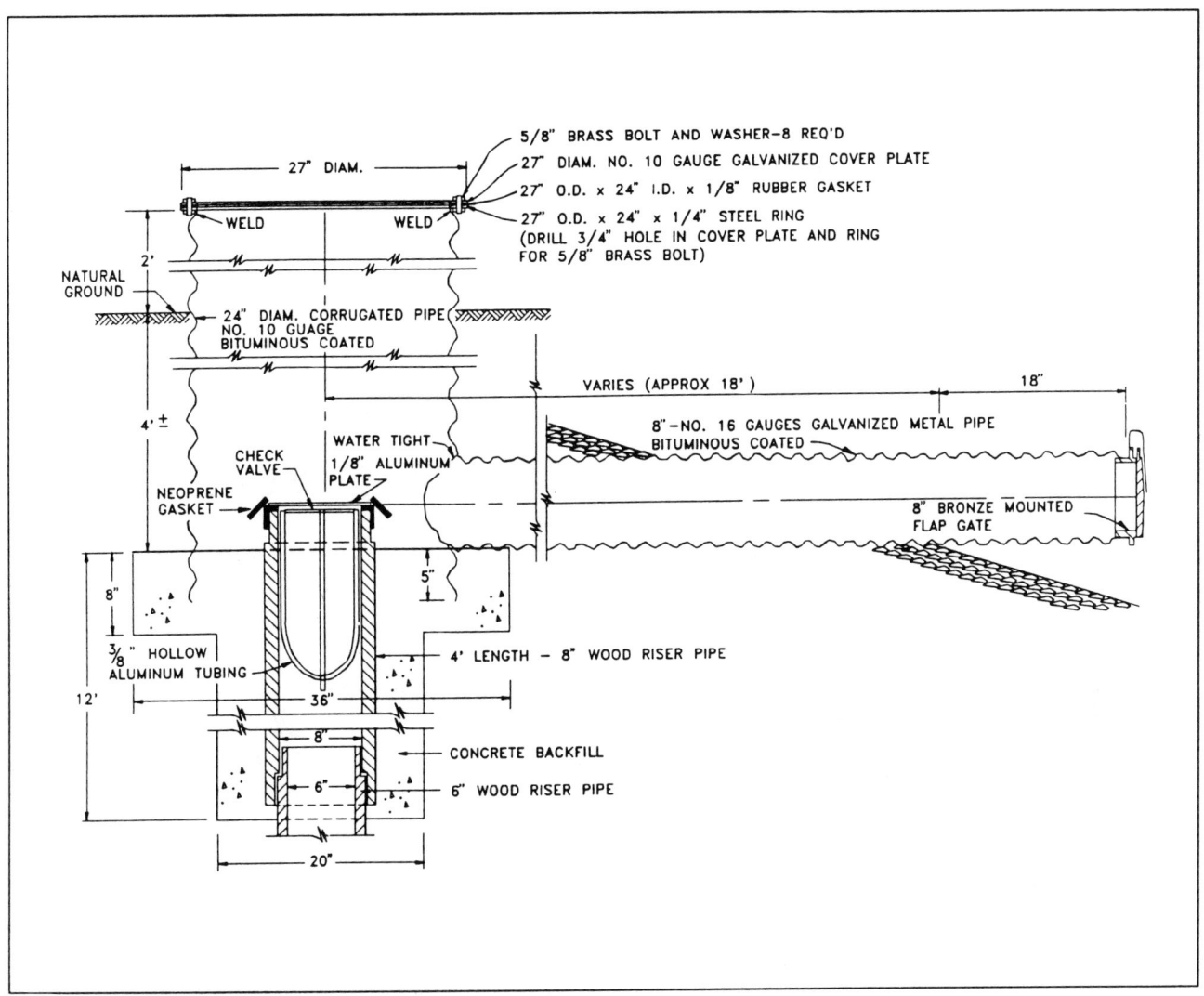

Figure 9-1. Typical detail of well top, check valve, and outlet

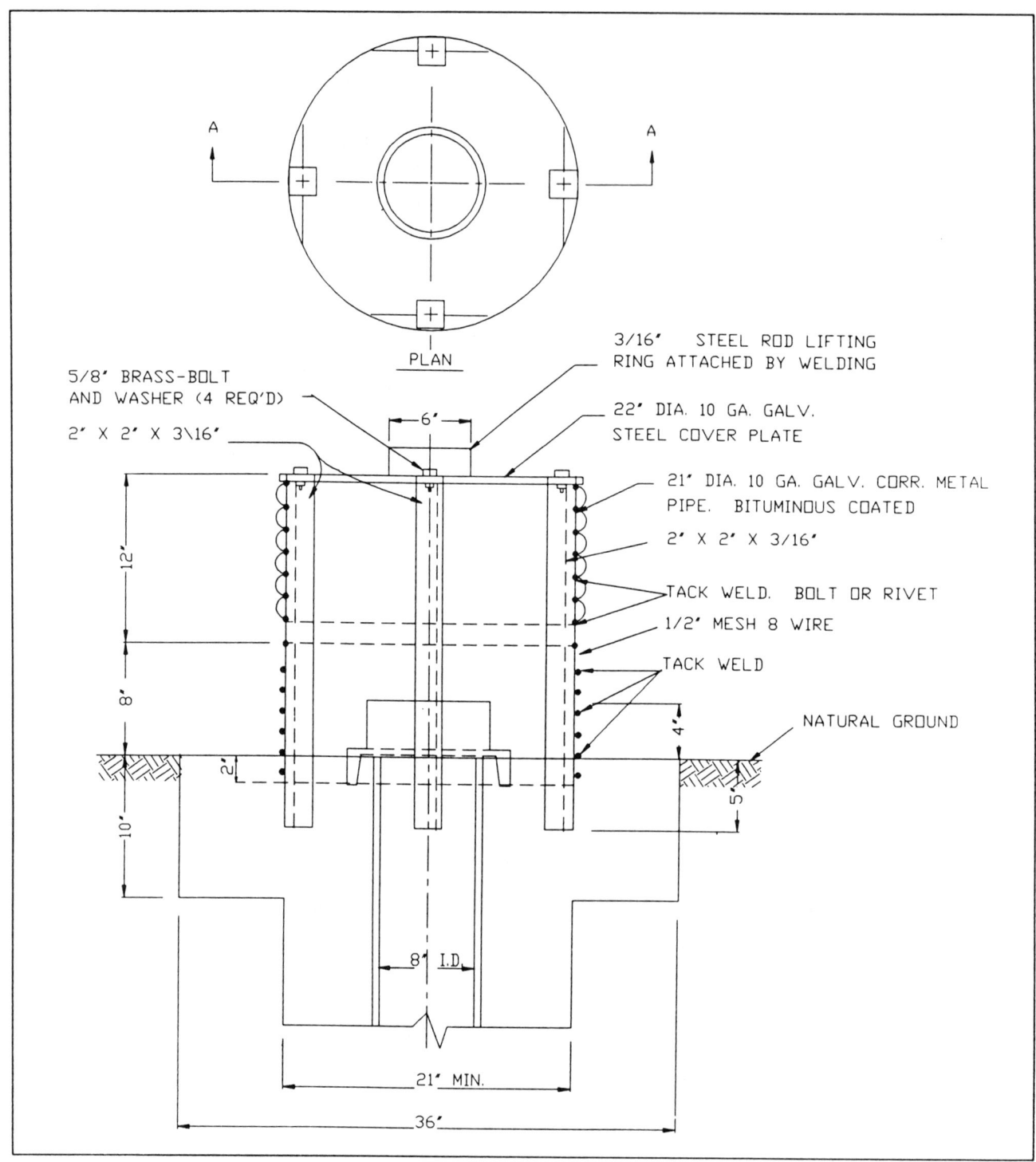

Figure 9-2. Metal well guard details

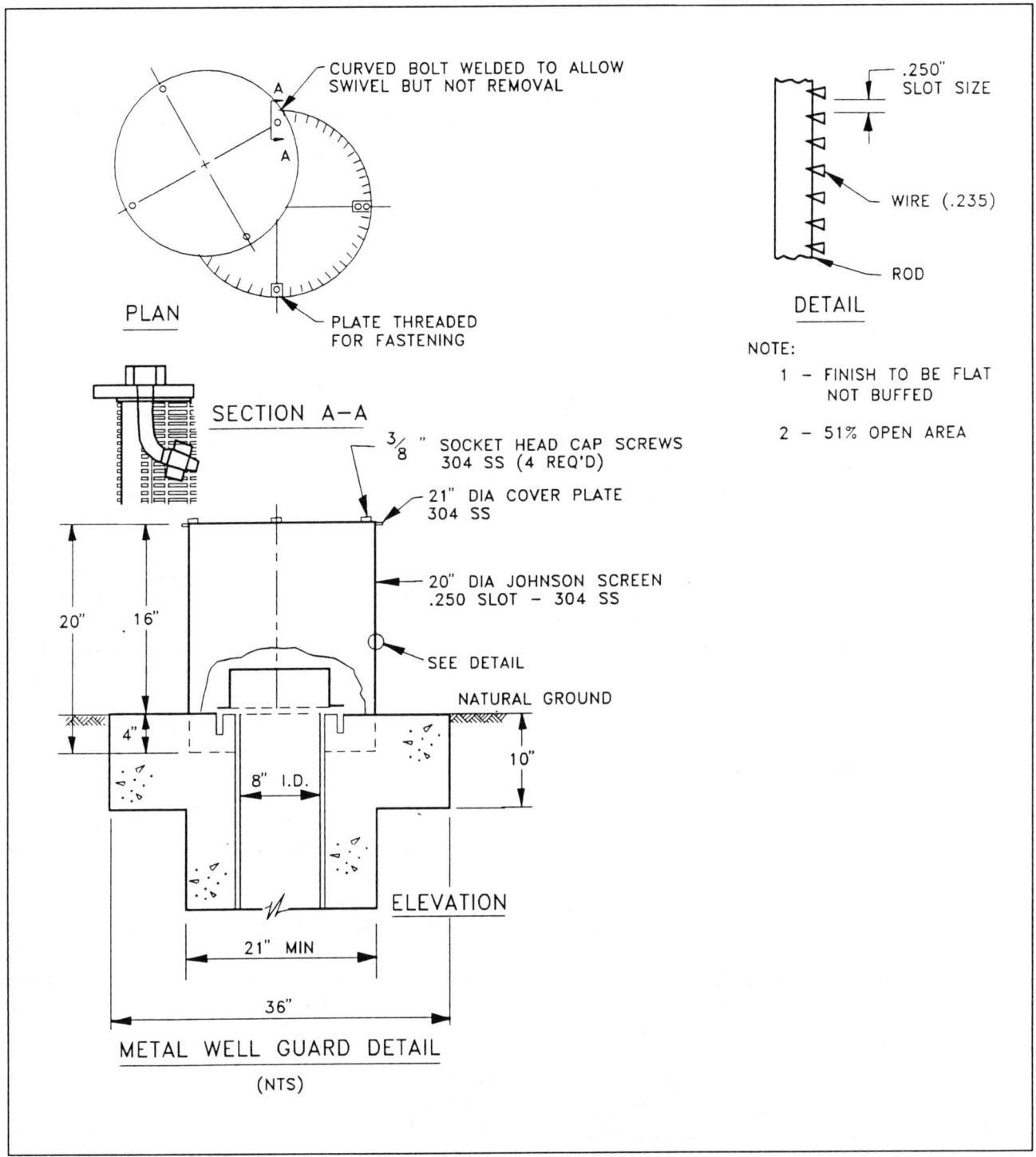

Figure 9-3. Alternative metal well guard.

CHAPTER 10

INSPECTION MAINTENANCE AND EVALUATION

10-1. General Maintenance

Relief wells require a certain amount of nominal maintenance to ensure their continued and proper functioning. Any trash or obstruction in the well or well guard should be removed immediately. Sand or other material that may have accumulated in and around flap gates to obstruct the flow or prevent functioning of the gates should be removed. Outfall ditches, bank slopes, or berms should be properly maintained in the vicinity of horizontal outlet pipes. The area in the immediate vicinity of the wells should be kept free from weeds, trash, and debris. Mowing and weed spraying should be extended at least 5 ft beyond the well, and the ground shaped and maintained for inspection and servicing of the wells.

10-2. Periodic Inspections

A. Periodic inspections of relief well systems should be carried out as described in ER 1110-2-100 and ER 110-2-1942. Relief well installations in readily accessible locations at dams and appurtenant structures where well flow is continuous should be visually inspected weekly. Observation should be made for evidence of wet spots on the dam or on the ground around the wells and structures, for evidence of sloughing or piping, for indications of discharge of sand or other materials from the wells, and for surficial signs of damage. The inspection should detect whether vandalism, theft, abuse by carelessness, unauthorized use of the wells or associated piezometers, or other irregularities have occurred. The inspection should include an examination of check valves, gaskets, well guards, cover plates, flap gates on tee outlets, and other appurtenances. Malfunctioning or damaged items should be repaired or replaced. At yearly intervals, piezometric levels and flow quantities should be measured, and wells should be sounded for evidence of deposition of sand or other material in the wells. Where relief wells penetrate two or more aquifers, the well flows at various depths should be checked at yearly intervals to determine whether flows between aquifers are occurring. Piezometric levels and flow quantities should also be measured approximately one week after the attainment of an unusually high reservoir level. Wells in relatively inaccessible locations, as beneath stilling basins, should be inspected whenever the structure is unwatered for a general maintenance inspection, or when there is evidence of significantly decreased effectiveness, as shown by changes in flow quantities or piezometric levels for a constant combination of reservoir level and tailwater level.

B. Flowing wells located in areas in which failure would not constitute a hazard to life or property, as on excavated slopes of canals, should be visually inspected at monthly intervals. Measurements of piezometric levels and flow quantities should be made annually.

C. Relief wells located along the toe of levees and at locations where they flow infrequently should be inspected annually, preferably immediately prior to normal high-water seasons and more often during major high waters. Flow quantities and piezometric levels should be measured approximately a week after a peak in the reservoir level or in the river level at a levee. Pumping tests should be performed at five-year intervals on wells that flow infrequently. The tests should be performed to determine the specific capacities and the efficiencies of the wells. The amount of sediment in the wells should be measured before and after performance of the pumping tests.

10-3. Pumping Tests

All wells should be pump tested every five years using procedures described in Chapter 8. Wells in relatively inaccessible locations should be pump tested whenever the structure is unwatered or when piezometric data indicate that well efficiency has decreased significantly. Wells should be checked for sanding before and after pump-

INSPECTION MAINTENANCE AND EVALUATION

ing. All wells requiring removal of sediment should be pump tested after cleaning to see if any appreciable loss of efficiency has resulted from foreign material entering the well. In the case of continuously flowing wells, the discharge should be measured at high pool levels to determine specific capacity and indications of sanding. If the pumping tests indicate that the specific capacity is less than 80 percent of that determined at the time of installation, then corrective measures should be employed. Investigations as described in Chapter 10 should be conducted prior to initiating the rehabilitation methods described in Chapter 11. If the rehabilitation methods are unsuccessful in restoring the wells to at least 80 percent of their original efficiency, consideration should be given to replacing these wells.

10-4. Records

A record should be kept of all inspections and maintenance performed on each well. The record should include all pumping test data, descriptions of rehabilitation efforts, and summaries of well flows and piezometric data during periods of high river stages or pool levels.

10-5. Evaluation

A. It should be noted that a reduction in well discharge accompanied by a fall in piezometric levels in downstream areas probably indicates a decrease in seepage due to siltation in the reservoir, riverbed areas, or riverside borrow pits, which is a favorable condition. It is possible, however, that such a reduction was caused by erosion or excavation of an impervious top stratum at a point downstream of the line of wells, thus permitting exit of seepage to tailwater much closer to the wells. This condition would be unfavorable, because it would indicate a higher value of the seepage gradient and an increased potential for piping immediately downstream from the well. A reduction in well discharge accompanied by an increase in piezometric levels indicates clogging or obstruction of the relief wells, and requires immediate remedial action. Observation of changes in flow and piezometric levels must be related to changes or lack of changes in both reservoir level and tailwater level. Often, variation in tailwater level at a dam has greater influence on well performance than variation in reservoir level, because the point at which the tailwater has access to the aquifer is considerably closer to the well than the point at which the reservoir pressure can enter the aquifer.

B. The values obtained from measurement or piezometric levels and flow quantities should be extrapolated to predict the values that would be produced by a maximum design reservoir or river elevation. If these values are greater than those for which the structure was designed, or if the specific capacities or the efficiencies of the wells are less than 80 percent of the values that were obtained at the time of installation of the wells, additional investigations should be performed to determine the cause of the inadequacies. Investigations may include the examination of the well screen by means of a borehole camera, sounding the well with a caliper, and the performance of chemical tests on the water and on any deposits or incrustations found in the well. If there are any inclinometer tubes installed in the foundation in the vicinity of the wells, they should be read to determine if there has been any horizontal movement of the foundation that would cause disruption of well screens or risers.

CHAPTER 11

MALFUNCTIONING OF WELLS AND REDUCTION IN EFFICIENCY

11-1. General

Relief wells may not function as intended and may also be subject to reduced efficiency with time. Failure of relief wells to function as intended can be attributed to a number of causes. Deficiencies in design can usually be assessed during initial operation of the well system. Based on piezometric and well flow data, an assessment of the effectiveness of the well system can be made and if considered inadequate, additional relief wells may be installed. Relief wells may malfunction for a variety of reasons including vandalism, breakage, or excessive deformation of the well screens due to ground movements, corrosion or erosion of the well screen, and a gradual loss in efficiency with time. The reduced efficiency generally determined as a percentage loss in specific capacity based on the specific capacity determined from pumping tests at the time of installation is a measure of increased well losses, which in turn result in higher landside heads. Thus, reduced well efficiency will result in hydrostatic heads larger than those anticipated in the design. The major causes of reduced specific capacity with time are mechanical, chemical, and biological.

11-2. Mechanical

Most relief wells undergo some loss in specific capacity probably due to the slow movement of foundation fines into the filter pack with a corresponding reduction in permeability. The process occurs more commonly in cases of poorly-designed filter packs, improper screen and filter pack placement, or insufficient well development. Generally, the major cause of reduced efficiency by mechanical processes is the introduction of fines into the well by backflooding of muddy surface waters. Normally, backflooding can be prevented by the use of check valves at the well outlet, however if not properly designed and maintained, the valves may not function as intended. The introduction of fines into the well and surrounding filter pack under backflow conditions can result in serious clogging which will result in reduced specific capacities.

11-3. Chemical

Chemical incrustation of the well screen, filter pack, and surrounding formation soils can be a major factor in specific capacity reduction with time. Chemical deposits forming within the screen openings reduce their effective open area and cause increased head losses. Deposits in the filter pack and surrounding soils reduce their permeability and also increase head losses. The occurrence of chemical incrustation is determined chiefly by water quality. The type and amount of dissolved minerals and gases in the water entering the well determine the tendency to deposit mineral matter as incrustations. The major forms of chemical incrustation include: (a) incrustation from precipitation of calcium and magnesium carbonates or their sulfates, and (b) incrustation from precipitation of iron and manganese compounds, primarily their hydroxides or hydrated oxides.

A. CAUSES OF CARBONATE INCRUSTATIONS. Chemical incrustation usually results from the precipitation of calcium carbonates from the ground water of the well. Calcium carbonate can be carried in solution in proportion to the amount of dissolved carbon dioxide in the ground water. For a well discharging from a confined aquifer, the hydrostatic pressure adjacent to the well is reduced to provide the gradient necessary for the well to flow. The reduction in pressure causes a release of carbon dioxide which in turn results in precipitation of some of the calcium carbonate. The precipitation tends to be concentrated at the well screen and surrounding filter pack where the maximum pressure reduction occurs. Magnesium bicarbonate may change to magnesium carbonate in the same manner, however incrustation from this source is seldom a problem as precipitation occurs only at very high levels of carbon concentration.

B. CAUSES OF IRON AND MANGANESE INCRUSTATION. Many ground waters contain iron and manganese ions if the pH is about 5 or less. Reduction of pressure due to well flow can disturb the chemical equilibrium of the ground water and result in the deposition of insoluble iron and manganese hydroxides. The hydroxides initially have the consistency of a gel, but eventually harden into scale deposits. Further oxidation of the hydroxides results in the formation of ferrous, ferric, or manganese oxides. Ferric oxide is a reddish brown deposit similar to rust, whereas the ferrous oxide has the consistency of a black sludge. Manganese oxide is usually black or dark brown in color. The iron and manganese deposits are usually found with calcium carbonate and magnesium carbonate scale.

11-4. Biological Incrustation

A. Iron bacteria are a major source of well screen and gravel pack contamination. They consist of organisms that have the ability to assimilate dissolved iron which the oxidize or reduce to ferrous or ferric ions for energy. The ions are precipitated as hydrated ferric hydroxide on or in their mucilaginous sheaths. The precipitation of the iron and rapid growth of the bacteria can quickly reduce well efficiency. Iron bacteria problems in ground water and wells are recognised throughout the world and are responsible for costly well maintenance and rehabilitation.

B. Despite the widespread familiarity with iron bacteria problems in wells, relatively little is known about their growth requirements. One reason for the lack of research on iron bacteria is that these organisms are difficult to culture for experimental study and that pure cultures of many of these organisms have never been obtained. Available information on the nature and occurrence of iron-precipitating bacteria in ground water is summarized by Hackett and Lehr in Leach and Taylor (1989).

C. In order to determine which genus of iron bacteria is contained in a particular water sample, a system of classification based on the physical form of these organisms has been employed by the water well industry (Driscoll 1986). The three general forms recognised are:

1. Siderocapsa. This organism consists of numerous short rods surrounded by a mucoid capsule. The deposit surrounding the capsule is hydrous ferric oxide, a rust-brown precipitate.

2. Gallionella. This organism is composed of twisted stalks or bands resembling a ribbon or chain. A bean-shaped bacterial cell, which is the only living part of the organism, is found at the end of the stalk.

3. Filamentous Group. This filamentous group consists of four genera: Chrenothrix, Sphaerotilus, Clonothrix, and Leptothrix. The organisms are structurally characterized by filaments which are composed of series of cells enclosed in a sheath. The sheaths are commonly covered with a slime layer. Both the sheath and slime layers or these organisms typically become encrusted with ferric hydrate resulting in large masses of filamentous growth and iron deposits.

D. IDENTIFICATION. The presence of iron bacteria is usually indicated by brownish red stains in well collector pipes or ditches. Television and photographic surveys can pinpoint the locations of screen incrustation, and samples of the incrustations can be obtained by a small bucket-shaped container. Samples can be sent to the USAE Waterways Experiment Station, or a private firm familiar with iron bacteria for identification. Identification is best accomplished by scanning electron or transmission electron microscopy and phase contrast techniques. Correct identification is necessary for selection of an appropriate treatment method.

E. PREVENTION. It is not clear whether iron bacteria exist in ground water before well construction takes place, or whether they are introduced into the aquifer from the foundation soils or in mix water during well construction. Evidence exists that iron bacteria may be carried from well to well on drill rods and other equipment and therefore every effort should be made to avoid introducing iron bacteria into a well during installation, maintenance, or rehabilitation operations. After completion of operations on a well, all drilling equipment, tools, bits and pumps, should be thoroughly disinfected by washing with a chlorine solution (100 ppm) before initiating work on another well.

CHAPTER 12

WELL REHABILITATION

12-1. General

The analysis of well discharge records and accompanying piezometric data will often indicate whether the relief wells are functioning as intended. A decrease in well discharges with time for similar pool or river stages with rising piezometric levels between wells is usually indicative of decreasing well efficiency. A quantitative measure of the loss in efficiency is only determined by carefully conducted pumping tests as previously described. Should the pumping tests indicate a reduction in specific capacity of more than 20 percent compared to that measured at installation, a detailed study should be made of the consequences of the reduction and what remedial measures should be employed. Generally, it may be possible to restore the wells to about their original efficiency by means of rehabilitation techniques.

Rapidly developing technology in the fields of chemistry and microbiology, as they are related to wells and aquifers, could negate portions of the following rehabilitation techniques, but the items covered are at least broadly covered and represent current practice. Environmental concerns (past and present chemical usage) also require that certain Federal, state, and local laws be followed and rehabilitation techniques may have to be modified to comply with these laws.

12-2. Mechanical Contamination

Plugging of relief wells by silts, clays, or other particulate media entering the filter pack either from the formation or through the top of the well is usually difficult to determine except as indicated by periodic pumping tests. If significant reductions in specific yield are noted, rehabilitation of the well is in order. Mechanical redevelopment of the well similar to that used to develop a new well should be the first step. Overpumping or pumping the well at the highest rate attainable is generally advantageous. Surging and the use of horizonal jetting devices also may produce beneficial results.

12-3. Chemical Treatment with Polyphosphates

Mechanical plugging of relief wells is corrected most often by chemical treatment with polyphosphates. These chemicals act as dispersing agents which causes silt and clay particles to repel one another and calcium, magnesium, and iron ions adhering to the particles to remain in a soluble state. The most widely used chemicals for this purpose are the glassy sodium phosphates which are inexpensive and readily available. The chemicals are usually applied in concentrations of 15 to 25 lb per 100 gal of water in combination with at least 50 ppm of chlorine (about one-half gal of 3 percent household bleach or chlorox in 100 gal of water). Phosphate solutions are mixed in a barrel or tank adjacent to the well. The material is best dissolved in small amounts in a wire basket or perforated container in agitated or swirling water. If the material is dropped directly into the tank or well, it will sink to the bottom and form a large gelatinous mass that could remain undissolved for some time. One of the most effective means of introducing the phosphate and chlorine solution into the well is by means of a horizontal jetting device. The well should then be surged vigorously prior to pumping. Three or more repetitions of injecting, surging, and pumping over a 2 to 4-hr cycle will be much more effective than a singe treatment with a longer detention time.

12-4. Chemical Incrustations

If the cause of reduced well efficiency is determined to be chemical incrustation, more frequent cleaning and maintenance should be initiated. If the efficiency remains low, consideration should be given to treating the well with a strong acid solution which can chemically dissolve the incrusting materials so that they can be pumped from the well. Acids most commonly used in well rehabilitation are hydrochloric acid, sulfamic acid, and hydroxyacetic (glycolic) acid. Acid treatment should be used with caution on wooden screen wells as the acid may tend to attack the lignin in

WELL REHABILITATION

the wood and cause severe damage. Methods for acid treatment of wells are described in detail by Driscoll (1986). The methods require great care and only experienced personnel with specialized equipment should be employed. Specialized firms with experience in this field should be utilized for this purpose.

12-5. Bacterial Incrustation

Incrustation of wells by iron bacteria is best controlled by a combination of chemical and physical treatments. Many chemical treatments have been suggested and applied in practice but their success has been variable as evidenced in many cases by recolonization or regrowth in the treated wells. A strong oxidizing agent such as chorine is widely used to limit the growth of iron bacteria. Chlorine, in the form of a gas, is used in the restoration of commercial wells; however safety and experience requirements limit its general application. A more convenient alternative is the use of hyperchlorite or other chlorine products (see Table 12-1). A discussion of procedures for the use of the various products is given by Driscoll (1986). Physical methods for control of iron bacteria are available, however sufficient research has not been accomplished to justify their use in relief wells. A survey of new techniques is presented by Hackett and Lehr in Leach and Taylor (1989).

12-6. Recommended Treatment

As clogging of well screens and filter materials is caused not only by the organic material produced by the bacteria, but also by oxides and hydroxides of iron and manganese, better results are usually obtained by treating the well alternately with a chlorine compound to attack the organic material and a strong acid to dissolve the mineral deposits. Between each treatment the well is pumped to waste to ensure that chlorine and acid are not in the well at the same time. A recommended procedure using the two procedures is:

A. Inject a mixture of acid, inhibitor, and wetting agent. The addition of a chelating agent such as hydroxyactic acid may sometimes be beneficial. An inhibitor is needed only if the well screen is metal. The amount of acid should be typically one and a half to two times the volume of the well screen. If a chelating agent is not used, iron will precipitate out if the pH rises above 3. The precipitate can result in clogging; therefore the pH should be monitored throughout the acid treatment and not be allowed to rise above 3 regardless of whether a chelating agent is used.

B. Gently agitate the solution with a jetting tool at 10-min intervals for a period of 1 to 2 hr.

C. Pump out a volume of solution equal to the volume of the well.

Table 12-1. Quantities of Various Chlorine Compounds Required to Provide as Much Available Chlorine as 1 lb of Chlorine Gas[1]

Chemical	% Available Chlorine	Number of lb Equivalent to 1 lb Cl_2
Chlorine Gas	100	1.0
Calcium Hypochlorite	65	1.54
Lithium Hypochlorite	36	2.78
Sodium Hypochlorite	12.5	8.0
Trichlorisocyanuric Acid[2]	90	1.11
Sodium Dichloroisocyanurate[2]	63	1.59
Potassium Dichloroisocyanurate[2]	60	1.67
Chlorine Dixoide	4	25.0
Chlorine Dioxide	2	50.0

Notes:
1. From Driscoll (1986). Reprinted with permission of Wheelabrator Engineered Systems, Inc.
2. Chlorine compounds that incorporate isocyanuric acid stabilize the chlorine against degradation from sunlight. Except for storage, the advantage offered by the addition of isocyanuric acid is less valuable in water wells.

D. Determine the pH of solution removed from the well. If the pH is more than 3, repeat steps **A** to **C**.

E. Allow the acid to remain in the well for a minimum of 12 hr and then pump to waste.

F. Inject a mixture of chlorine and one or more chloric-stable surfactants (detergents and wetting agents, for example). The concentration of the chlorine should exceed 1,000 ppm.

G. Gently agitate the solution with a jetting tool at 10-min intervals every 2 hr for the first 8 hr and then at 8 hr intervals for at least 24 hr.

H. Pump out a volume of solution equal to the volume of the well.

I. Determine chlorine concentration. If the concentration is less than 10 percent of the original concentration, repeat steps **F** to **H**.

J. Perform a pumping test on the well. If the specific capacity has improved by more than 5 percent, repeat the entire procedure until the specific capacity does not improve by 5 percent.

12-7. Specialized Treatment

The USAE Waterways Experiment Station personnel, funded under a repair evaluation maintenance and restoration (REMR) work unit, developed a field procedure (Kissane and Leach 1991) for cleaning water wells that provides initial kill of the active bacteria in the well, dissolves the biomass in the screen, in the gravel pack, and some distance into the aquifer, and provides some inhibition of future growth. The procedure was developed using a patented process known as the Alford Rodgers Cullimore Concept (ARCC). The procedures in general include an initial well diagnosis performed with a prepackaged field microbiological test kit which is designed to give a qualitative indication of the types of bacterial and chemical agents at work in the wells, and a very general indication of the bacterial concentrations. The initial water chemistry is also measured prior to treatment. A treatment is then designed with the information from the tests, targeting the problematic agents with an appropriate set of chemicals. Redevelopment of the wells using the ARCC method is based on the use of blended chemicals and high temperature (BCHT) and is divided into three principle elements of treatment:

A. SHOCK. This phase is achieved by adding high temperature chlorinated water to the well and surrounding aquifer to "shock" kill or reduce the impact of deleterious algae and bacteria. The water is chlorinated to >700 ppm with gaseous chlorine to avoid binders found in powdered chlorine and is applied to the well as steam until the well temperature is brought above 120 deg F for massive bacteria kill. The chlorine treatment remains in the well for a specified period of time; mechanical surging is used; and pumping follows for removal of the initial loosened biomass.

B. DISRUPT. This phase is achieved by the addition of chemical agents, acids and surfactants, and steam to the well and surrounding aquifer while the well is pressurized. Mechanical surging to break up organic and mineral clogging in the system is also used. The mechanical surging and chemical set time are important during this phase to achieve dissolution of the remaining biomass.

C. DISPERSE. This phase of treatment consists of removal of the material that has been clogging the well and aquifer. Acceptance criteria for the well are checked and further cycles are considered or a final cold chlorination treatment is applied for inhibition of any remaining bacterial colonies.

APPENDIX A

REFERENCES

A-1 Required Publications

1. TM 5-818-5 Dewatering and Ground Water Control (Nov 83)
2. ER 1110-2-100 Periodic Inspection and Continuing Evaluation of Completed Civil Works Structures (Apr 88)
3. ER 1110-2-110 Instrumentation for Safety-Evaluations of Civil Works Projects (Jul 85)
4. ER 1110-2-1942 Inspection, Monitoring, and Maintenance of Relief Wells (Feb 88)
5. EM 1110-1-1804 Geotechnical Investigations (Feb 84)
6. EM 1110-2-1602 Hydraulic Design of Reservoir Outlet Works (Oct 80)
7. EM 1110-2-1901 Seepage Analysis and Control for Dams (Sep 86)
8. EM 1110-2-1905 Design of Finite Relief Well Systems (Mar 63)
9. EM 1110-2-1906 Laboratory Soils Testing Ch 1-2 (Nov 70)
10. EM 1110-2-1907 Soil Sampling (Mar 72)
11. EM 1110-2-1908 Instrumentation of Earth and Rock-Fill Dams—Part 1 (Aug 71), Part 2 (Nov 76)
12. EM 1110-2-1913 Design and Construction of Levees (Mar 78)
13. EM 1110-2-2300 Earth and Rock-Fill Dams General Design and Construction Considerations (May 82)

A-2 Related Publications

1. Banks, D. C. 1963 (Mar) "Three-Dimensional Electrical Analogy Seepage Model Studies; Appendix A: Flow to Circular Well Arrays Centered Inside a Circular Source, Series G," Technical Report No. 3-619, US Army Engineer Waterways Experiment Station Vicksburg, MS.
2. Banks, D. C. 1965 (Mar). "Three-Dimensional Electrical Analogy Seepage Model Studies; Appendix B: Flow to a Single Well Centered Inside a Circular Source, Series H." Technical Report No. 3-619, US Army Engineer Waterways Experiment Station, Vicksburg, MS.
3. Barron, R. A. 1948. "The Effect of a Slightly Pervious Top Blanket on the Performance of Relief Wells," *Proceedings of the Second International Conference on Soil Mechanics and Foundation Engineering*, Rotterdam, Netherlands, Vol 4, p. 342
4. Barron, R. A. 1982 (Sept). "Mathematical Theory of Partially Penetrating Relief Wells," Unpublished report prepared for the US Army Engineering Waterways Experiment Station, Vicksburg, MS.
5. Bennett, P. T., and Barron, R. A. 1957. "Design Data for Partially Penetrating Relief Wells," *Proceedings of the Fourth International Conference on Soil Mechanics and Foundation Engineering*, London, United Kingdom, Vol II.
6. Conroy, P. 1984. "Computer Program for Relief Well Design According to TM 3-424," US Army Engineer District, St. Louis, St. Louis, MO.
7. Cunny, R. W., Agostinelli, V. M., Jr., and Taylor, H. M., Jr. 1989 (Sept). "Levee Underseepage Software User Manual and Validation," Technical Report REMR-GT-13, US Army Engineer Waterways Experiment Station, Vicksburg, MS.
8. Dachler, A. E. 1936. "Grundwasserstromung," Julius Springer, Vienna.
9. Driscoll, F. G. 1986. "Groundwater and Wells" 2d., Johnson Division, SES, Inc., St Paul, MN.
10. Duncan J. M. 1963 (Mar). "Three-Dimensional Electrical Analogy Seepage Model Studies," Technical Report No. 3-619, US Army Engineer Waterways Experiment Station, Vicksburg, MS.
11. Environmental Protection Agency. 1976 (Jul), "Manual of Methods for Chemical Analysis of Water and Wastes," EPA-625-6-76-003a. Available from Environmental Protection Agency, 26 W. St. Clair Street, Cincinnati, OH 45268.

12. Ferris, J. G., Knowles, D. B., Brown, R. H., and Stellman, R. W. 1962. "Theory of Aquifer Tests," Geological Survey Water-Supply Paper 1536-E, US Government Printing Office, Washington, DC.

13. Forcheimer, P. H. 1914. *Hydraulik,* Teubner, Berlin.

14. Freeze, R. A. and Cherry, J. A. 1976. *Groundwater,* Prentice-Hall, Englewood Cliffs, N.J.

15. Hadj-Hamou, T., Tavassoli, M., and Sherman, W. C. 1990 (Sept). "Laboratoy Testing of Filters and Slot Sizes for Relief Wells," *Journal of Geotechnical Engineering,* American Society of Civil Engineers, Vol. 116, No. 9.

16. Harr, M. E. 1962. *Groundwater and Seepage,* McGraw-Hill, New York.

17. Johnson, S. J. 1947. Discussion of "Relief Wells for Dams and Levees," by Middlebrooks, T. A., and Jervis, W. H. *Transactions of the American Society of Civil Engineers,* Vol 112.

18. Kissane, J., Leach, R. 1991 (Mar). "Redevelopment of Relief Wells, Upper Wood River Drainage and Levee District, Madison County, Illinois," Technical Report REMR GT-16, US Army Engineer Waterways Experiment Station, Vicksburg, MS.

19. Kozeny, J. 1993. "Theorie and Berechnvng der Brunnen," Wasserkroft W. Wasser Wirtschoft, Vol 29.

20. Leach, R. E. and Taylor H. M. Jr. 1989 (Jul). "Proceedings of REMR Workshop on Research Priorities for Drainage System and Relief Well Problems," Final Report, US Army Engineer Waterways Experiment Station, Vicksburg MS.

21. Mansur, C. I., and Dietrich, R. J. 1965 (Jul)."Pumping Test to Determine Permeability Ratio," *Journal of the Soil Mechanics and Foundation Division,* American Society of Civil Engineers, Vol. 91, No. SM4.

22. Mansur, C. I., and Kaufman, R. I. 1955 (May). "Control of Underseepage, Mississippi River Levees, St. Louis District, Corps of Engineers," US Army Engineer Waterways Experiment Station, Vicksburg, MS.

23. Mansur, C. I., and Kaufman, R. I. 1962. "Dewatering," *Foundation Engineering,* edited by G. A. Leonards, McGraw-Hill, New York.

24. McAnear, C. L., and Trahan, C. C. 1972 (Jan). "Three-Dimensional Seepage Model Study, Oakley Dam, Sangamon River, Illinois," Miscellaneous Paper S-72-3, US Army Engineer Waterways Experiment Station, Vicksburg, MS.

25. Middlebrooks, T. A. 1948. "Seepage Control for Large Earth Dams." *Transactions of the Third International Congress on Large Dams,* Stockholm, Sweden.

25. Middlebrooks, T. A. and Jervis, W. H. 1947. "Relief Wells for Dams and Levees," *Transactions of the American Society of Civil Engineers,* Vol. 112.

26. Montgomery, R. L. 1972 (Jun). "Investigation of Relief Wells, Mississippi River Levees, Alton to Gale, Illinois," Miscellaneous Paper S-72-21, US Army Engineer Waterways Experiment Station, Vicksburg, MS.

27. Moser, J. H. and Huibregtse, K. R. 1976 (Sept). "Handbook for Sampling and Sample Preservation of Water And Wastewater," EPA-600-4-76-049; Available from Environmental Protection Agency, 26 W. St. Clair Street, Cincinnati, OH 45268.

28. Muskat, M. 1937. *The Flow of Homogeneous Fluids Through Porous Media,* J. W. Edwards, Ann Arbor, MI.

29. Todd. D. K. 1980. *Groundwater Hydrology,* 2d ed., John Wiley & Sons, New York.

30. Turnbull, W. J. and Mansur, C. I. 1954. "Relief Well Systems for Dams and Levees," *Transactions of the American Society of Civil Engineers,* Vol 119, Paper No. 2701.

31. US Army Engineer District, Albuquerque. 1967 (Jul). "Cochiti Lake, Rio Grande, New Mexico, Embankment and Conveyance Channel," Design Memorandum No. 9, Albuquerque, NM.

32. US Army Engineer Waterways Experiment Station. 1956 (Oct). "Investigation of Underseepage and Its Control, Lower Mississippi River Levees," 2 Vols, Technical Memorandum No. 3-424, Vicksburg, MS.

33. US Army Engineer Waterways Experiment Station. 1958 (Jun). "Analysis of Piezometer and Relief Well Data, Arkabutla Dam," Technical Report No. 3-479, Vicksburg, MS.

34. US Army Engineer Waterways Experiment Station. 1973. "Computer Program H2030-Discharge in Pressure Conduits Using the Darcy-Weisbach Formula," Vicksburg, MS.

35. Warriner, J. B., and Banks, D. C. 1977. "Numerical Analysis of Partially Penetrating Well Arrays," Technical Report S-77-5, US Army Engineer Waterways Experiment Station, Vicksburg, MS.

APPENDIX B

Mathematical Analysis of Underseepage and Substratum Pressure

B-1. General

The design of seepage control measures for levees and dams often requires an underseepage analysis without the use of piezometric data and seepage measurements. Contained within this appendix are equations by which an estimate of seepage flow and substratum pressures can be made, provided soil conditions at the site are reasonably well defined. The equations contained herein were developed during a study by the US Army Engineer Waterways Experiment Station (1956)[1] of piezometric data and seepage measurements along the Lower Mississippi River and confirmed by model studies. The following discussion is presented in terms of levee underseepage, however the analyses and equations are considered equally applicable to dam foundations. It should be emphasized that the accuracy obtained from the use of equations is dependent upon the applicability of the equation to the condition being analyzed, the uniformity of soil conditions, and the evaluation of the various factors involved. As is normally the case, sound engineering judgement must be exercised in determining soil profiles and soil input parameters for these analyses.

B-2. Assumptions

It is necessary to make certain simplifying assumptions before making any theoretical seepage analysis. The following is a list of such assumptions and criteria necessary to the analysis set forth in this appendix:

A. Seepage may enter the previous substratum at any point in the foreshore (usually at riverside borrow pits) and/or through the riverside top stratum.

B. Flow through the top stratum is vertical.

C. Flow through the pervious substratum is horizontal. Flow in the vertical direction is entirely disregarded.

D. The levee (including impervious or thick berm) and the portion of the top stratum beneath it are impervious.

E. All seepage is laminar.

In addition to the above, it is also required that the foundation be generalized into a pervious sand or gravel stratum with a uniform thickness and permeability and a semipervious or impervious top stratum with a uniform thickness and permeability (although the thickness and permeability of the riverside and landside top stratum may be different).

B-3. Factors Involved in Seepage Analyses

The volume of seepage (Q_s) that will pass beneath a levee and the artesian pressure that can develop under and landward of a levee during a sustained high water are related to the basic factors given and defined in Table B-1 and shown graphically in Figure B-1. Other terms used in the analyses are defined as they are discussed in subsequent paragraphs.

B-4. Determination of Factors Involved in Seepage Analyses

Many of the factors necessary to perform a seepage analysis, such as exploration and testing, have previously been mentioned in the text, however they are discussed in more detail as they apply to each specific factor. The use of piezometric data, although rarely available on new projects, is mentioned primarily because it is

---

[1] References cited in this appendix are listed in Appendix A.

Table B-1. Examples of Transformation Procedure

Strata	Actual Thickness ft	Actual Permeability cm/sec	$F_t = \dfrac{k_b}{k_n}$	Transformed Thickness, ft for $k_b = 1 \times 10^{-4}$ cm/sec
Clay	5	1×10^{-4}	1	5.0
Sandy Silt	8	2×10^{-4}	1/2	4.0
Silty Sand	5	10×10^{-4}	1/10	0.5
	$Z = \overline{18}$			$Z_b = \overline{9.5}$

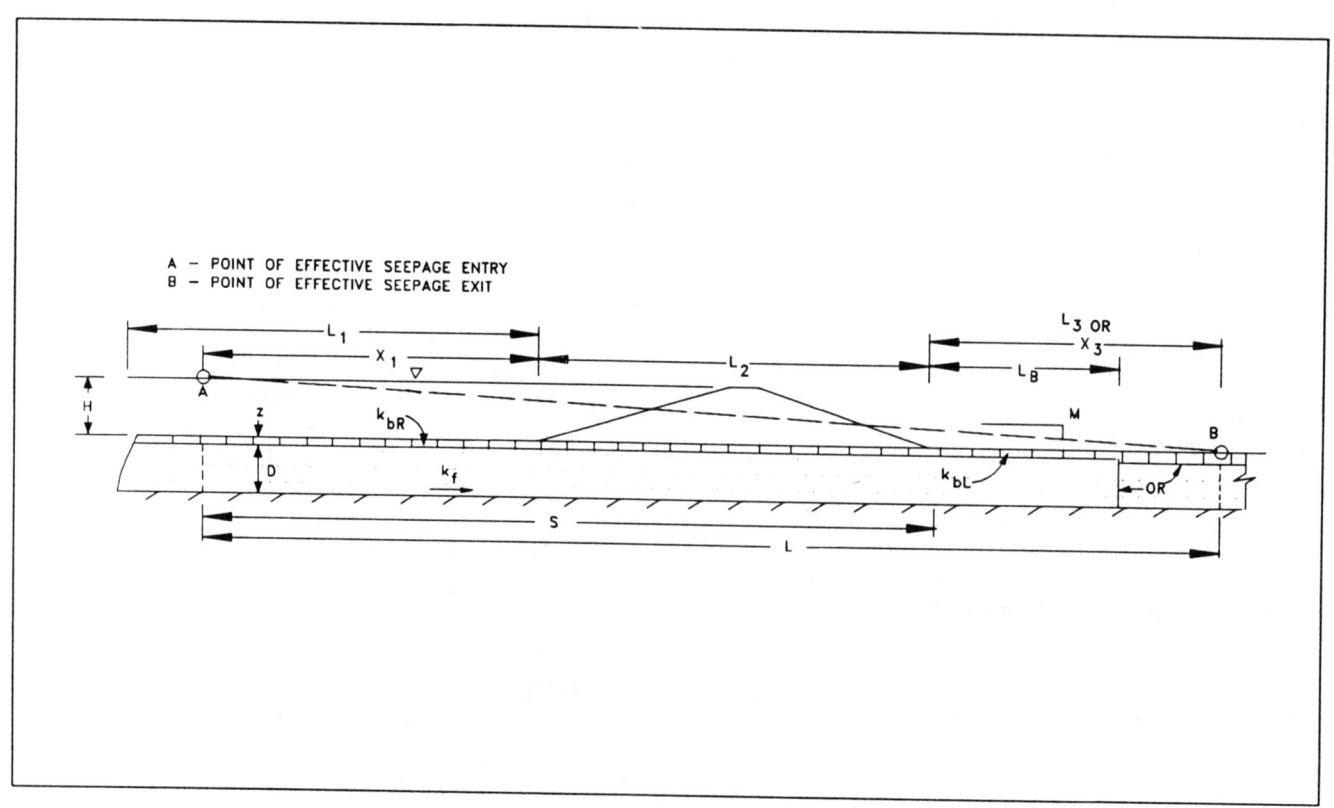

Figure B-1. Illustration of symbols used in Appendix B

MATHEMATICAL ANALYSIS OF UNDERSEEPAGE AND SUBSTRATUM PRESSURE

not infrequent for seepage analyses to be performed as a part of remedial measures for existing levees in which case piezometric data often are available.

A. NET HEAD H. The net head on a levee is the height of water on the riverside above the tailwater or natural ground surface on the landside of the levee. H is usually based on the net levee grade but is sometimes based on the design or project floodstage.

B. THICKNESS, Z_t AND VERTICAL PERMEABILITY, K_b, OF TOP STRATUM. Where the thickness of the riverside blanket differs from that of the landside blanket, the designations, Z_{bR} and Z_{bL} are used. Similarly the permeability of the riverside and landside blankets are designated k_{bR} and k_{bL}, respectively.

1. Exploration. The thickness of the top stratum, both riverward and landward of the levee, is extremely important in a seepage analysis. Exploration to determine this thickness usually consists of auger borings with samples taken at 3- to 5-ft intervals and at every change of material. Boring spacing will depend on the potential severity of the underseepage problem, but should be laid out for sampling the basic geologic features with intermediate borings for check purposes. Landside borings should be sufficient to delineate any significant geological features as far as 500 ft away from the levee toe. The effect of ditches and borrow areas must be considered.

2. Transformation. The top stratum in most areas is seldom composed of one uniform material but rather usually consists of several layers of different soils. If the in situ vertical permeability of each soil layer (k_n) is known, it is possible to transform the top stratum to an equivalent stratum of effective thickness and vertical permeability. However, a reasonably accurate seepage analysis can also be made by assuming a uniform vertical permeability for the top stratum equal to the permeability of the most impervious strata and then using the transformation factor given in Equation B-1 to determine a transformed thickness for the entire top stratum.

$$F_t = \frac{k_b}{k_n} \quad (B-1)$$

where F_t is the transformation factor. Some examples using this procedure are given in Table B-1 and in Figure B-1. A generalized top stratum having a uniform vertical permeability of 1×10^{-4} cm/sec and thickness of 9.5 ft would then be used in the seepage analysis for computation of effective blanket lengths. However, the thickness Z_{bL} may or may not be the effective thickness of the landside top stratum Z_t that should be used in determining the hydraulic gradient through the top stratum and the allowable pressure beneath the top stratum. The transformed thickness or the top stratum equals the in situ thicknesses of all strata above the base of the least pervious stratum plus the transformed thicknesses of the underlying more pervious top strata. Thus, Z_{bL} will equal Z_t only when the least pervious stratum is at the ground surface. Several examples of this transformation are given in Figure B-2. To make the final determination of the effective thicknesses and permeabilities of the top stratum, conditions at least 200 to 300 ft landward of the levee must be considered. In addition, certain averaging assumptions are almost always required where soil conditions are reasonably similar. Existing landward conditions or critical areas should be given considerable weight in arriving at such averages.

C. THICKNESS D AND PERMEABILITY k_f OF PERVIOUS SUBSTRATUM. The thickness of the pervious substratum is defined as the thickness of the principal seepage-carrying stratum below the top stratum and above rock or other impervious base stratum. It is usually determined by means of deep borings although a combination of shallow borings and seismic or electrical resistivity surveys may also be employed. The thickness of any individual pervious strata within the principal seepage-carrying stratum must be obtained by deep borings. The average horizontal permeability k_f of the pervious substratum can be determined by means of a field pump test on a fully penetrating well as described in the main text. For areas where such correlations exist, their use will usually result in a more accurate permeability determination than that from laboratory permeability tests. In addition to the methods above, if the total amount of seepage passing beneath the levee (Q_s) and the hydraulic grade line beneath the levee (M) are known, k_f can be estimated from the equation

$$k_f = \frac{Q_s}{M} \quad (B-2)$$

D. DISTANCE FROM RIVERSIDE LEVEE TOE TO RIVER, L_1. This distance can usually be estimated from topographic maps.

E. BASE WIDTH OF LEVEE AND BERM, L_2. The distance, L_2, can be determined from

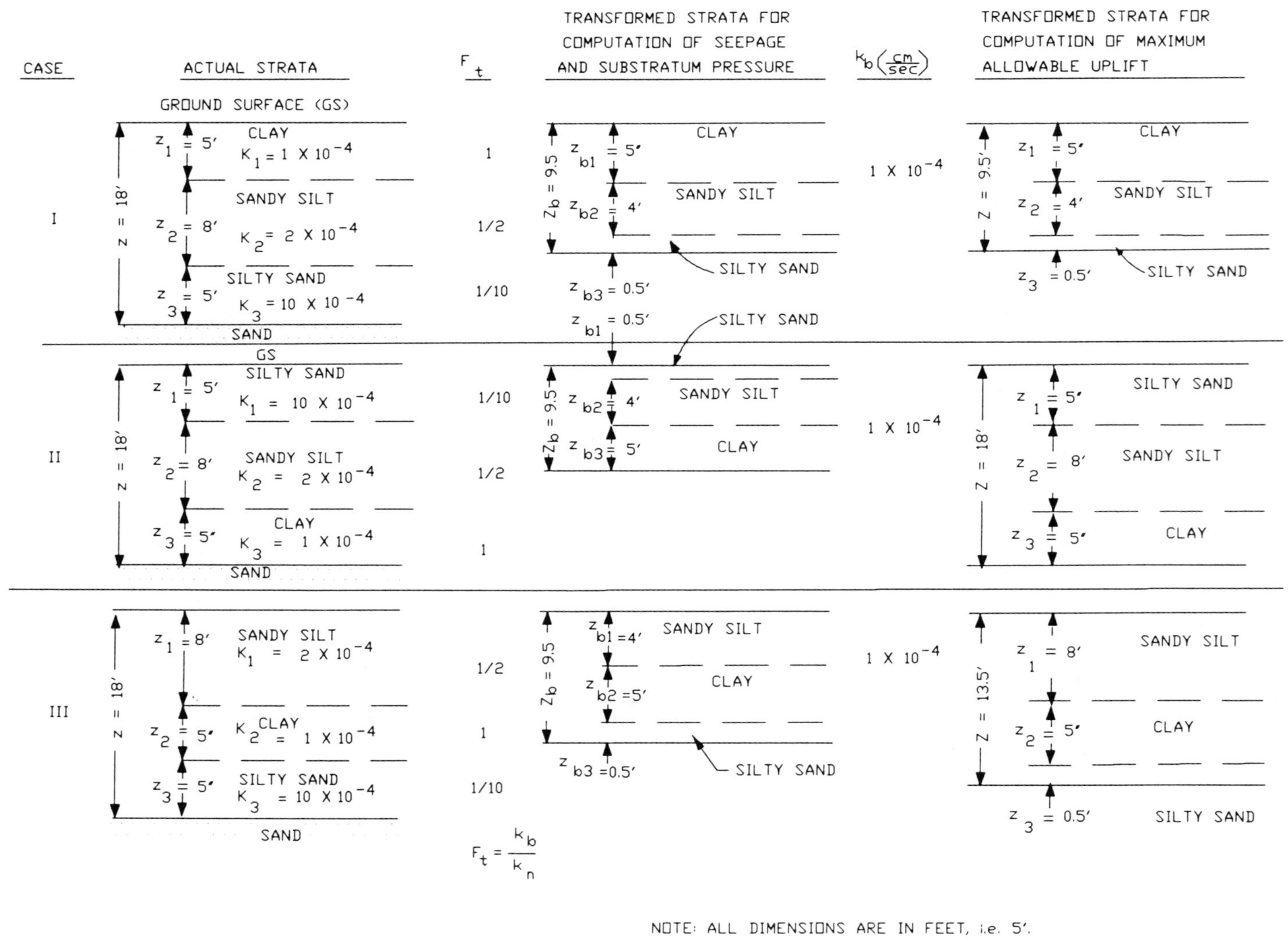

Figure B-2. Transformation of top strata

anticipated dimensions of new levees or by measurement in the case of existing levees.

F. LENGTH OF TOP STRATUM LANDWARD OF LEVEE TOE, L_3.

This distance can usually be determined from borings, topographic maps, and/or field reconnaissance. To determine this distance, careful consideration must be given to any geological feature that may affect the seepage analysis. Of special importance are deposits of impervious materials, such as clay plugs which can serve as seepage barriers. If the barrier is located near the landside toe, it could force the emergence of seepage at the near edge and have a pronounced effect on the seepage analysis.

G. DISTANCE FROM LANDSIDE LEVEE TOE TO EFFECTIVE SEEPAGE EXIT, X_3.

The effective seepage exit (Point B, Figure B-1) is defined as that point where a hypothetical open drainage face would result in the same hydrostatic pressure at the landside levee toe and would cause the same amount of seepage to pass beneath the levee as would occur for actual conditions. Point B is located where the hydraulic grade line beneath the levee projected landward with a slope M intersects the ground water or tailwater. If the length of foundation and top stratum beyond the landside levee toe L_3 is known, x_3 can be estimated from the following equations:

(1) For $L_3 = \infty$

$$x_c = \frac{1}{c} = \sqrt{\frac{k_f Z_{bL} D}{k_{bL}}} \qquad \text{(B-3)}$$

where

$$c = \sqrt{\frac{k_{bL}}{k_f Z_{bL} D}} \qquad \text{(B-4)}$$

(2) For L_B = finite distance to a seepage block

$$x_3 = \frac{1}{c \tanh cL_b} \qquad \text{(B-5)}$$

(3) For L_3 = finite distance to an open seepage exit

$$x_3 = \frac{\tanh cL_3}{c} \qquad \text{(B-6)}$$

H. DISTANCE FROM EFFECTIVE SOURCE SEEPAGE ENTRY TO RIVERSIDE LEVEE TOE, X_1.

The effective source of seepage entry into the pervious substratum (Point A, Figure B-1) is defined as that line riverward of the levee where a hypothetical open seepage entry face fully penetrates the pervious substratum. An impervious top stratum between the seepage entry and the levee would produce the same flow and hydrostatic pressure beneath and landward of the levee as would occur for the actual conditions riverward of the levee. Effective seepage entry is also defined as that line or point where the hydraulic grade line beneath the levee projected riverward with a slope, M, intersects the river stage.

1. If the distance to the river from the riverside levee toe, L_1, is known, and no riverside borrow pits or seepage blocks exist, x_1 can be estimated from the following equation:

$$x_1 = \frac{\tanh cL_1}{c} \qquad \text{(B-7)}$$

where C is calculated from Equation B-4 using appropriate properties of the riverside top stratum.

2. If a seepage block (usually a wide, thick deposit of clay) exists between the riverside levee toe and the river in order to prevent any seepage entrance into the pervious foundation beyond that point, x_1 can be estimated from the following equation:

$$x_1 = \frac{1}{c \tanh cL_1} \qquad \text{(B-8)}$$

where L_1 equals distance from riverside levee toe to seepage block and c is calculated from Equation B-4.

3. The entrance conditions often are such that an assumption of a vertical entrance face is not reasonable. Two limiting cases are shown in Figure B-3. The additional effective length, ΔL_1, may be obtained for either Case A which assumes a uniformly sloping entrance face or Case B which assumes a combined infinite horizontal entrance face with a vertical entrance face, D', varying from O to D (see Figure B-3).

I. Critical gradient for landside top stratum, i_c. The critical gradient is defined as the gradient required to cause boils or heaving (flotation) of the landside top stratum and is taken

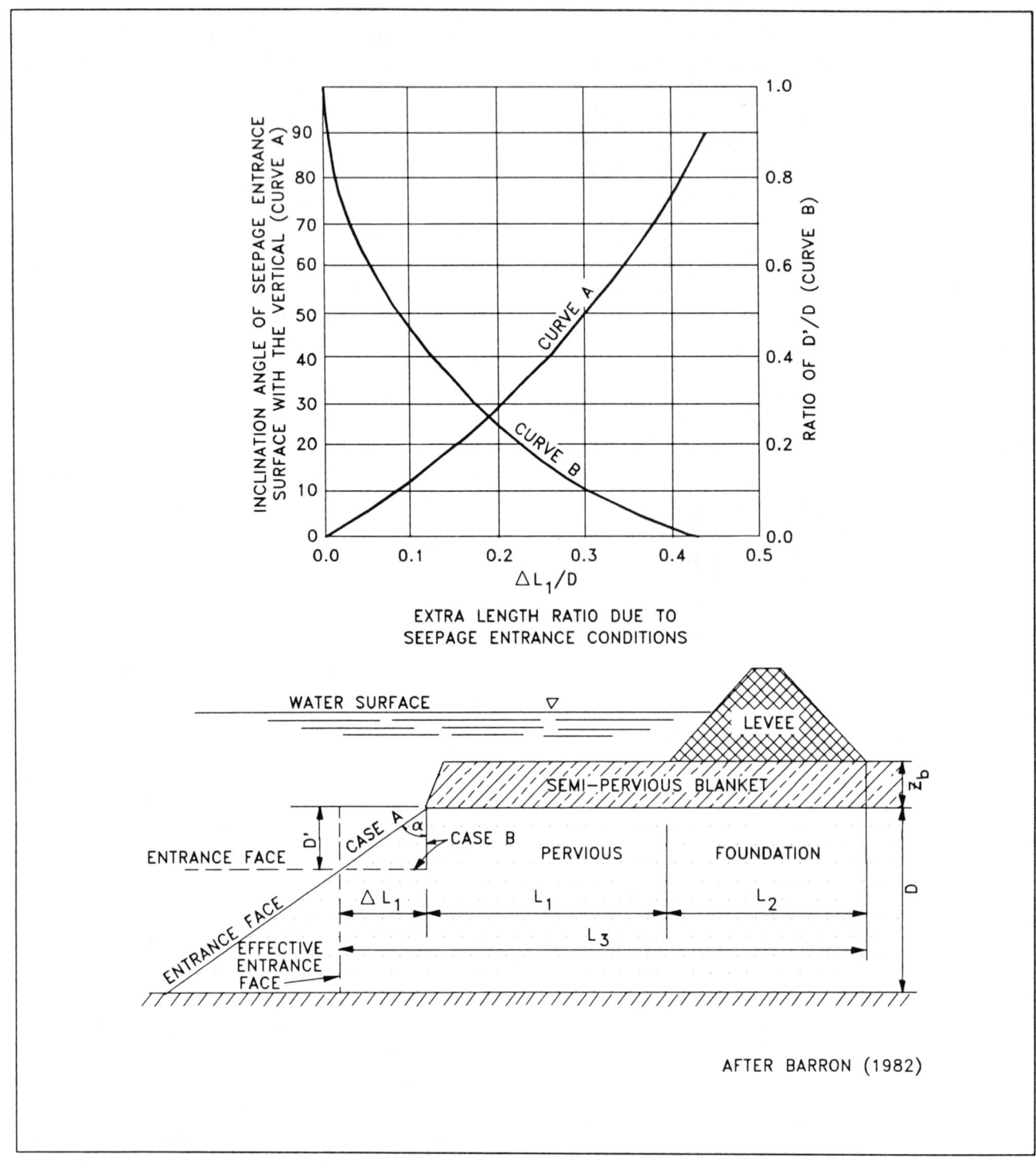

Figure B-3. Corrections for nonvertical entrance face (after Barron 1982)

as the ratio of the submerged or buoyant unit weight of soil, γ', comprising the top stratum and the unit weight of water, γ_w, or

$$i_c = \frac{\gamma'}{\gamma_w} = \frac{G_s - 1}{1 + e} \qquad \text{(B-9)}$$

where

G_s = specific gravity of soil solids
e = void ratio

J. SLOPE OF HYDRAULIC GRADE LINE BENEATH LEVEE, M. The slope of the hydraulic grade line in the pervious substratum beneath a levee can best be determined from readings of piezometers located beneath the levee where the seepage flow lines are essentially horizontal and the equipotential lines vertical. The slope of the hydraulic grade line cannot be reliably determined, however, until the flow conditions have developed beneath the levee. If no piezometer readings are available, as in the case for new levee design, M must be determined by first establishing the effective seepage entrance and exit points and then connecting these points with a straight line, the slope of which is M.

B-5. Computation of Seepage Flow and Substratum Hydrostatic Pressures

A. GENERAL

1. Seepage. For a levee underlain by a pervious foundation, the natural seepage per unit length of levee, Q_s, can be expressed by

$$Q_s = \varphi k_f D \qquad \text{(B-10)}$$

were φ is the shape factor. This equation is valid provided the assumptions upon which Darcy's law is based are met. The mathematical expressions for the shape factor φ (subsequently given in this appendix) depend upon the dimensions of the generalized cross section of the levee and foundation, the characteristics of the top stratum both riverward and landward of the levee, and the pervious substratum. Where the hydraulic grade line M is known from piezometer readings, the quantity of underseepage can be determined from

$$Q_s = M k_f D \qquad \text{(B-11)}$$

2. Excess hydrostatic head beneath the landside top stratum. The excess hydrostatic head h_o beneath the top stratum at the landside levee toe is related to the net head on the levee, the dimensions of the levee and foundation, permeability of the foundation, and the character of the top stratum both riverward and landward of the levee. The head h_x beneath the top stratum at a distance x landward from the landside levee toe can be expressed as a function of the net head H and the distance x, although it is more conveniently related to the head h_o at the levee toe. When h_x is expressed in terms of h_o it depends only upon the type and thickness of the top stratum and pervious foundation landward of the levee; the ratio h_x/h_o is thus independent of riverward conditions. Expressions for φ, h_o and h_x for various boundary conditions are presented below.

B. CASE 1—NO TOP STRATUM. Where a levee is founded directly on pervious materials and no top stratum exists either riverward or landward of the levee (Figure B-4a), the seepage Q_s can be obtained from Equation B-12. The excess hydrostatic head landward of the levee is zero and $h_o = h_x = 0$. The severity of such a condition in nature is governed by the exit gradient and seepage velocity that develop at the landside levee toe which can be estimated from a flow net compatible with the value of S computed from Equation B-12.

C. CASE 2—IMPERVIOUS TOP STRATUM BOTH RIVERSIDE AND LANDSIDE. This case is found in nature where the levee is founded on thick (15-ft) deposits of clay or silts with clay strata. For such a condition, little or no seepage can occur through the landside top stratum.

1. If L_3 is infinite in landward extent or the pervious substratum is blocked landward of the levee, no seepage occurs beneath the levee and $Q_s = 0$. The head beneath the levee and the landside top stratum is equal to the net head on the levee at all points so that $H = h_o = h_x$.

2. If an open seepage exists in the impervious top stratum at some distance L_3 from the landside toe (i.e. L_3 is not infinite) as shown in Figure B-4b, the distance from the feet toe of the levee to the effective seepage entry (river, borrow pit, etc.) is $L_1 = L_2$. The equation for the shape factor is given by Equation B-13, and the heads h_o and h_x can be computed from Equations B-14 and B-15 respectively.

D. CASE 3—IMPERVIOUS RIVERSIDE TOP STRATUM AND NO LANDSIDE TOP STRATUM. This case is shown in Figure B-4c. The condition may occur naturally or where extensive landside borrowing has taken place resulting in

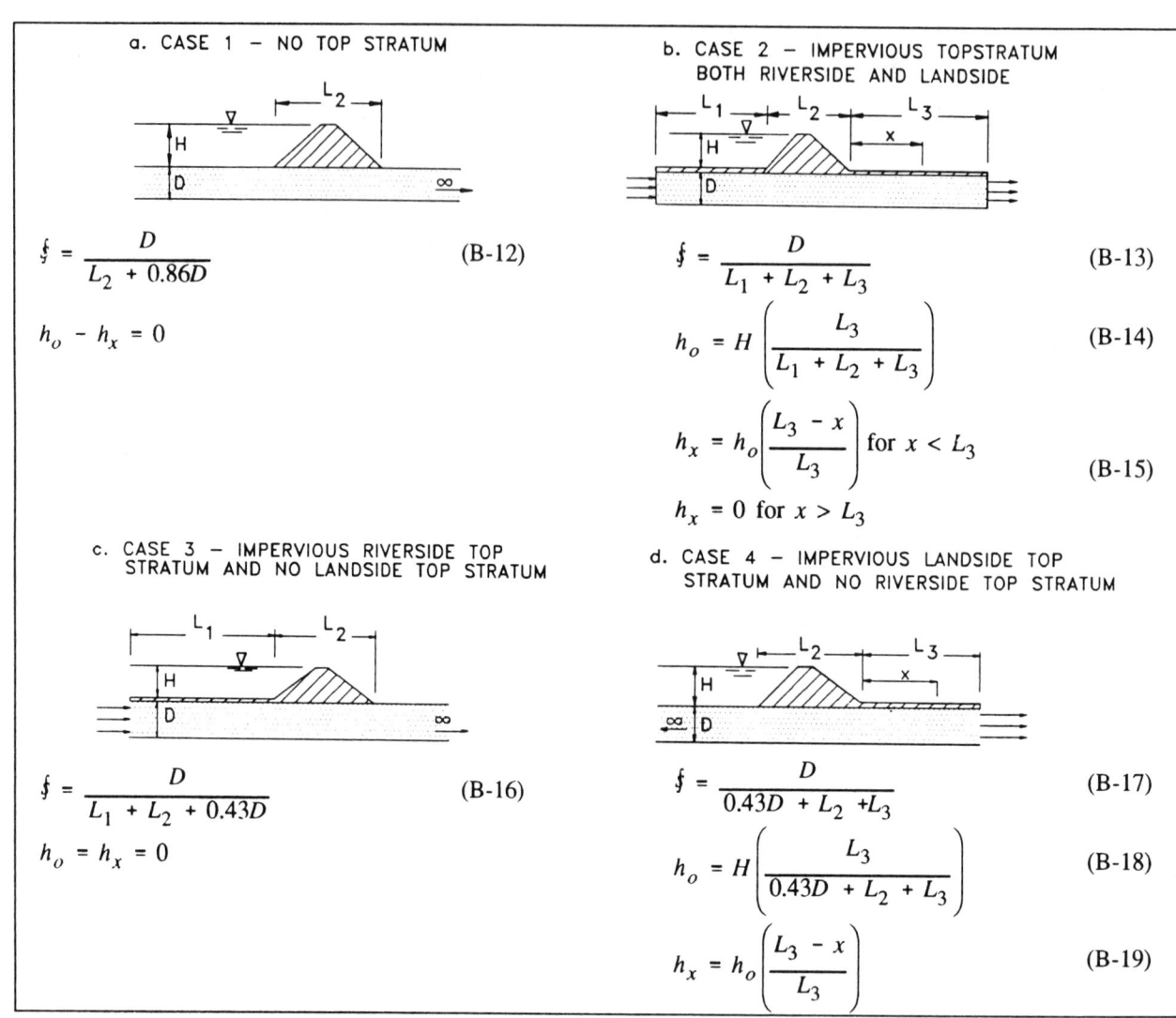

Figure B-4. Equations for computation of underseepage flow and substratum pressures for Cases 1 through 4

removal of all impervious material landward of the levee for a considerable distance. The shape factor is computed from Equation B-16. The excess head at the top of the sand landward of the levee is zero, and the danger from piping must be evaluated from the upward gradient obtained from a flow net.

E. CASE 4—IMPERVIOUS LANDSIDE TOP STRATUM AND NO RIVERSIDE TOP STRATUM. This case is more common than Case 3 and occurs when extensive riverside borrowing has resulted in removal of the riverside impervious top stratum (Figure B-4d). For this condition, the shape factor is computed from Equation B-17; the heads h_o and h_x are computed from Equations B-18 and B-19, respectively.

F. CASE 5—SEMIPERVIOUS RIVERSIDE TOP STRATUM AND NO LANDSIDE TOP STRATUM. This case is illustrated in Figure B-5a. The same equation for the shape factor as was used in Case 3 can be applied to this condition provided x_1 is substituted for L_1 in Equation B-16 resulting in Equation B-20. Since no landside top stratum exists, $h_o = h_x = 0$.

G. CASE 6—SEMIPERVIOUS LANDSIDE TOP STRATUM AND NO RIVERSIDE TOP STRATUM. This case is illustrated in Figure B-5b. The shape factor is given by Equation B-21 and the heads h_o and h_x are computed from Equations B-22 and B-23 respectively.

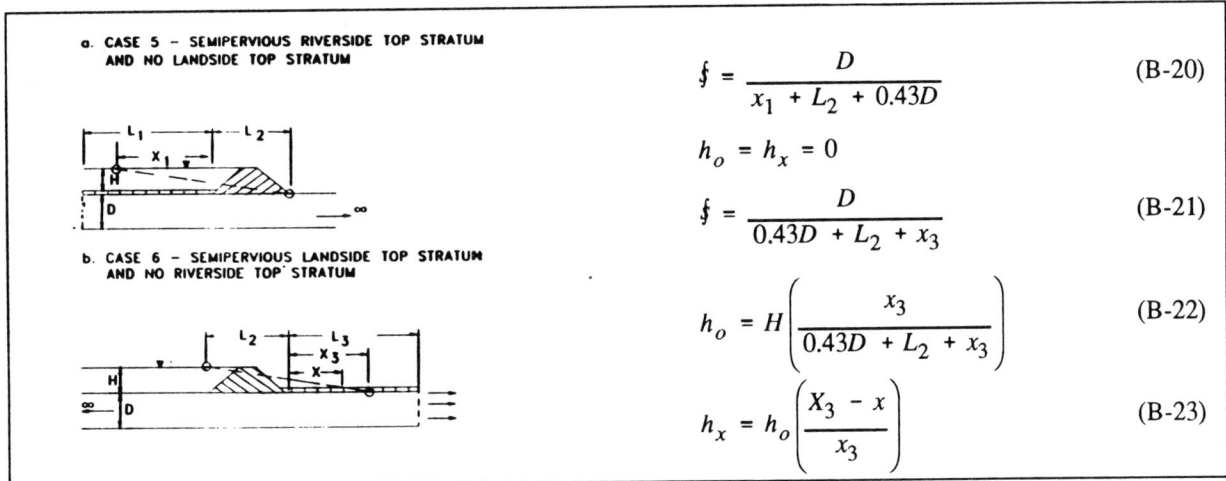

a. CASE 5 – SEMIPERVIOUS RIVERSIDE TOP STRATUM AND NO LANDSIDE TOP STRATUM

$$\mathcal{s} = \frac{D}{x_1 + L_2 + 0.43D} \qquad \text{(B-20)}$$

$$h_o = h_x = 0$$

b. CASE 6 – SEMIPERVIOUS LANDSIDE TOP STRATUM AND NO RIVERSIDE TOP STRATUM

$$\mathcal{s} = \frac{D}{0.43D + L_2 + x_3} \qquad \text{(B-21)}$$

$$h_o = H\left(\frac{x_3}{0.43D + L_2 + x_3}\right) \qquad \text{(B-22)}$$

$$h_x = h_o\left(\frac{X_3 - x}{x_3}\right) \qquad \text{(B-23)}$$

Figure B-5. Equations for computation of underseepage flow and substratum pressures for Cases 5 and 6

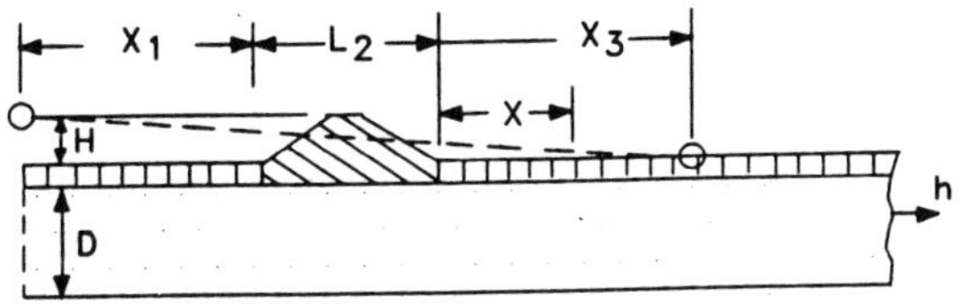

a. $L_B = \infty$

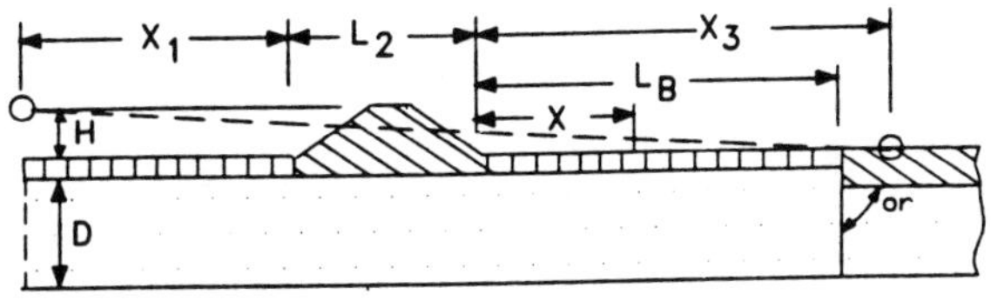

$$\mathcal{s} = \frac{D}{x_1 + L_2 + x_3} \qquad \text{(B-24)}$$

$$h_o = \frac{H x_3}{x_1 + L_2 + x_3} \qquad \text{(B-25)}$$

$$c = \frac{k_{bL}}{k_f z_{bL} D}$$

$$h_x = h_o\, e^{-cx} \qquad \text{(B-26)}$$

b. L_B IS FINITE TO A SEEPAGE BLOCK

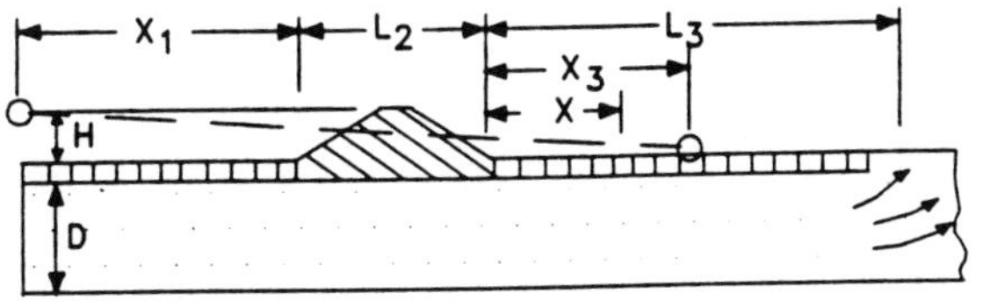

$$h_x = h_o \frac{\cosh c(L_B - x)}{\cosh c L_B} \qquad \text{(B-27)}$$

$$h_x @ L_3 = \frac{h_o}{\cosh c L_B} \qquad \text{(B-28)}$$

c. L_3 IS FINITE TO AN OPEN SEEPAGE EXIT

$$H_x = h_o \frac{\sinh c(L_3 - x)}{\sinh c L_3} \qquad \text{(B-29)}$$

$$H_x @ L_3 = 0 \qquad \text{(B-30)}$$

Figure B-6. Equations for computation of underseepage and substratum pressures for Case 7

H. CASE 7—SEMIPERVIOUS TOP STRATA BOTH RIVERSIDE AND LANDSIDE.

Where both the riverside and landside top strata exist and are semipervious (Figure B-6), the shape factor can be computed from Equation B-24. The head h_o is given by Equation B-25. The head h_x beneath the semipervious top stratum depends not only on the head h_o, but also on conditions landward of the levee and can be computed from Equations B-26 through B-30.

APPENDIX C

LIST OF SYMBOLS

a	Well-spacing	h_o	Excess hydrostatic head beneath the top stratum at landside levee toe
a,b	Dimensions defining boundary of non-circular source	h_p	Net head at any point p
A_c	Effective average distance from well center to external boundary of source	h_w	Head at well
		h_x	Head beneath top stratum at distance x from landside toe of levee
A_e	Equivalent radius of a group of wells	h_{av}	Average net head in plane of wells corrected for well losses
c	(1) Conversion factor used in determining the effective length of pervious foundation covered by a semipervious blanket	h_{wj}	Head at well j in a system of n wells
		Δh_d	Difference in elevation between landside piezometric surface and well outlets
C	(2) Hazen-Williams coefficient	H	Net head on well system. Difference between riverside pool and landside tailwater
C	(3) One-half the length of a finite line source		
C_u	Coefficient of uniformity	H_l	Total head measured from bottom of pervious foundation
d	Thickness of a pervious stratum	H_e	Entrance loss in screen and filter
$\bar{d}$	Transformed thickness of pervious stratum layer with thickness $= d$	H_f	Frictional head loss
		H_m	Net head midway between wells
d_m	Thickness of pervious foundation layers (summation of $m = 1$ to $m = n$)	H_v	Velocity head loss
		H_w	Well losses
D	Thickness of pervious foundation	H_{av}	Average net head in plane of wells
$\bar{D}$	Transformed thickness of pervious foundation	H_{m_n}	Net head midway between n number of wells
D'	Thickness of sloping entrance face of pervious foundation	H_{m_∞}	Net head midway between wells in infinite line
D_n	Grain size for which n percent of the sample is smaller	ΔH_m	Excess head above the well outlet midway between wells
e	Void ratio	i_c	Critical hydraulic gradient
F_t	Permeability transformation factor	i_o	Downward force acting against uplift pressure
FS	Factor of safety		
g	Acceleration due to gravity	k	Coefficient of permeability
G_p	Flow correction factor for partially-penetrating well	$\bar{k}$	Coefficient of permeability of transformed stratum
$G(\bar{T})$	A function used in analysis of partially-penetrating well	k_b	Vertical permeability of top stratum
		k_e	Effective permeability of multistratum transformed aquifer
h	Net head on well system corrected for well losses		
h_a	Allowable net head beneath top stratum at landside toe of levee on dam	k_f	Horizontal permeability of pervious substratum
h_c	Piezometric head midway between wells in circular array	k_h	Coefficient of permeability in horizontal direction
h_d	Maximum head landward from a slot or line of wells	k_L	Coefficient of permeability from laboratory tests
h_g	Head at boundary between artesian and gravity flow	k_v	Coefficient of permeability in vertical direction
h_m	Net head midway between well corrected for well losses		

k_m	Permeability of pervious foundation layers (summation of $m = 1$ to $m = n$)	r_{wj}	Effective well radius of well j
		R	Radius of influence
k_n	Vertical permeability of individual layers comprising top stratum (n = layer number)	R_i	Radius of influence of ith well
		R_j	Radius of influence of well j
k_{bL}	Vertical permeability of landside top stratum	S	Distance from effective seepage entry to line of wells
k_{bR}	Vertical permeability of riverside top stratum	S_i	Distance from infinite line seepage to multiple wells
L	Distance from source to seepage exit	v	Flow velocity in well
L_1	Distance from source to landside toe of levee or dam	W	Actual well penetration
		$\overline{W}$	Effective well penetration
L_2	Base width of impervious levee and berm	x,y,z	Cartesian coordinates
L_3	Length of pervious foundation and top stratum beyond landside toe of levee	x_a	Distance from effective seepage entry to point where gravity flow occurs
L_B	Distance from line of wells to blocked exit	x_g	Length over which gravity flow occurs to well line
L_e	Distance from line of wells to seepage exit	x_1	Distance from effective seepage entry to riverside toe of levee
$\Delta L1$	Extra length of pervious foundation due to sloping entrance face	x_3	Distance from landside toe of levee to effective seepage exit
n	Number of wells in group	Z	Thickness of top stratum
M	Slope of hydraulic grade line (at middepth of pervious foundation)	Z_b	Transformed thickness of top stratum
		Z_c	Thickness of top stratum below collector ditch
ΔM	Difference in slopes of hydraulic grade line riverside and landside of toe of levee	Z_n	Thickness of individual layers comprising top stratum (n = layer number)
		Z_t	Transformed thickness of landside top stratum for uplift computations
p	A point		
Q_a	Artesian component of seepage flow	Z_{bL}	Transformed thickness of landside top stratum
Q_g	Gravity component of seepage flow		
Q_s	Total amount of seepage beneath levee	Z_{bR}	Transformed thickness of riverside top stratum
Q_w	Discharge from a single well		
Q_{wj}	Flow from well j	α	Angle of entrance face
Q_{wp}	Flow from partially penetrating well	γ	Unit weight of soil
Q_{sw}	Seepage flow beyond well system	γ_w	Unit weight of water
		γ'	Submerged unit weight of soil
r	Radial distance from well (distance from point p to real well)	Θ_a	Average uplift factor
		Θ_m	Midwell uplift factor
r_c	Rows of circular array of well	$\Delta\Theta$	Change in Θ_a and Θ_m per 1 log cycle of a/r_w
r_i	Distance from i^{th} well to point p		
r_{ij}	Distance from well i to well j		
r'	Distance from point p to image well	δ	Offset distance between well and center of circular source
r_o	Distance from well to center of finite line source		
r_w	Radius of well	φ	Shape factor

INDEX

Abandoned wells 59
Allowable heads 11, 12
Anisotropic conditions 6
Applications, relief wells 2-4
Arkabutla Dam, Mississippi 4
Average uplift factor 23, 29

Backfilling 57-58
Backflooding 60, 66
Bailing and casing drilling method 51
Biological incrustation 67, 69-70
Blocked exit 25-26, 33
Boundary conditions 42
Bucket augers 51

Check valves 60, 61
Chemical composition, ground water 6, 11
Chemical development 54
Chemical incrustation 66-67, 68-69
Chemical treatment 68
Chlorine solution 51, 53
Circular array of wells 22
Circular source, seepage 13, 22
Clogging 66, 68; treatment of 68, 69-70
Complex boundary conditions 15
Computer programs 45
Costs 46

Design heads 42
Design procedures 42; infinite line of wells 42-45
Design, wells 35-41; computer programs 45; well systems 42-48
Discharge, below ground surface 26
Disinfection 51, 53
Drilling equipment 49, 50, 51
Drilling methods 49-51

Effective well radius 41
Effective well penetration 18, 20
Entrance losses 38, 39

Factor of safety 11, 12
Filter gradation 36-37

Filter placement 53-54
Finite line of wells, head distribution 45-46
Finite line source, seepage 15
Finite well lines 28
Flow changes, investigations of 65
Forcheimer equation 21
Fort Peck Dam, Montana 2
Foundation investigations 6; permeability 6, 10
Friction losses 39-41

Ground water, chemical composition 6, 11

Head, determination of 21-22
Head distribution 45-46
Head losses 38; 42
History 2, 4
Hydrostatic pressures, control of 2

Infinite barrier 15
Infinite line sink 15
Infinite line of wells, design procedure 42-45
Infinite line of wells 22, 24-27, 32, 33, 34
Infinite line source, seepage 13, 22, 32
Inspections 64-65
Installation 49-59; problems 49, 50-51
Iron bacteria 67, 69

Maintenance 64-65
Materials 35; selection of 35-36
Mechanical development 54-57
Metal well guards 60, 61-62
Method of multiple images 22
Method of images 15
Midwell uplift factor 23, 30
Multiple well systems 21-34

Noncircular source, seepage 13

Optimum design 46, 48
Outlets 60-63; protection of 60; structures 46

Partially penetrating wells 15, 18, 19
Performance evaluation 57
Plastic sleeves 60
Plugging 66, 68; treatment of 68, 69-70
Pumping, inspection 64-65; well development 55-57; well testing 57

Records 59, 65
Reduction of specific capacity 66-67
Rehabilitation 68-70
Relief wells, definition of 1; description of 2, 3
Reverse-rotary drilling method 49-51
Riser pipes, installation of 53

Sand infiltration 57
Screen-opening size 37
Seepage 46-47; analysis 11, 73-82; source of 13-15
Seepage control measures 2
Seepage flow computation 79-82
Single wells, analysis of 13-20
Spillways 46

Standard rotary drilling method 49
Sterilization 58-59
Substratum hydrostatic pressure, analysis of 73-82
Substratum hydrostatic pressures computation 79-82
Surging block 54-55

Testing equipment 57
Top stratum, impervious 22, 24-27, 28
Transformation procedure 75, 76
Tremie pipes 53-54, 58

Velocity head losses 41
Visual inspections 64

Water jetting 54, 55
Well description 35
Well factors 23-24
Well losses 38-41
Well screens, installation of 53; types of 36